A Workbook of Clinical Chemistry

Case Presentation and
Data Interpretation

A *Workbook of Clinical Chemistry*

Case Presentation and Data Interpretation

Companion to

Clinical Chemistry in Diagnosis and Treatment

Philip D Mayne

MD (Dublin), BA (Mod), MSc (London), FRCPI, FRCPath, FFPath (RCPI)

Consultant in Chemical Pathology, The Children's Hospital, Temple Street, The Rotunda Hospital and Our Lady's Hospital for Sick Children, Dublin; Formerly Senior Lecturer in Chemical Pathology, Charing Cross and Westminster Medical School, The Westminster Hospital, London

Andrew P Day

MA (Cambridge), MSc, MB, BS (London), MRCPath

Senior Registrar in Chemical Pathology, The Royal Infirmary, Bristol; Formerly Lecturer in Chemical Pathology, Charing Cross and Westminster Medical School, The Westminster Hospital, London

Edward Arnold
A member of the Hodder Headline Group
LONDON BOSTON MELBOURNE AUCKLAND

First published in Great Britain 1994 by
Edward Arnold, a division of Hodder Headline PLC,
338 Euston Road, London NW1 3BH

Whilst the advice and information in this book is believed to be true and
accurate at the date of going to press, neither the authors nor the publisher
can accept any legal responsibility or liability for any errors or omissions
that may be made. In particular (but without limiting the generality of the
preceding disclaimer) every effort has been made to check drug dosages;
however it is still possible that errors have been missed. Furthermore,
dosage schedules are constantly being revised and new side effects
recognized. For these reasons the reader is strongly urged to consult the
drug companies' printed instructions before administering any of the drugs
recommended in this book.

British Library Cataloguing in Publication Data
A catalogue record for this book is available from the British Library

ISBN 0 340 57646 4 ✓

1 2 3 4 5 94 95 96 97 98

Typeset in 11/12 pt Garamond by
Scribe Design, Gillingham, Kent, UK.
Printed and bound in Great Britain by
Butler and Tanner Ltd, Frome, Somerset, UK.

Contents

Tables

Foreword

We are flattered to have been asked to write a foreword to this book.

These reports of genuine cases bring together relevant information on each patient, and so illustrate the clinical nature of chemical pathology (clinical chemistry) and of other so-called 'support services'. Photographs of patients, X-rays and the findings of other pathology departments are included and reinforce the message stated in the Introduction, based on a chapter in the companion book *Clinical Chemistry in Diagnosis and Treatment*. It is in the patient's best interest that no subject, whether primarily ward or clinic based, or originating in a pathology or other department, should be considered in isolation. All are equally 'clinical' and teamwork involving staff in each is the basis of good modern medical practice. Exchange of ideas between specialists in every field is essential if errors, sometimes dangerous, in selection of tests and interpretation of results, and in treatment based on them, are to be minimized. The cases recorded in this book illustrate this point well.

Tables summarizing causes of biochemical abnormalities and the section concentrating on 'data interpretation' will be useful as revision for examinations. Examinations are by no means the end of learning: we are students all our lives and must use our own, continually expanding, experience to distinguish between the common and the rare, and our own reasoning to test hypotheses. No book or lecture can be more than a guide, and the information given in it must continually be tested against the reader's own knowledge and conclusions. This book allows us to do so. We are sure that it will be deservedly successful.

Joan F. Zilva
Peter R. Pannall
May 1994

Preface

This book, designed to partner the sixth edition of *Clinical Chemistry in Diagnosis and Treatment*, should help the student interpret biochemistry data in context. It thus reinforces the view, always stressed in the parent book, that clinical chemistry is a clinical subject.

Of the 15 sections the first and last (and several tables and figures) are closely based on the text of successive editions of that book, first by Zilva and Pannall, then by Zilva, Pannall and Mayne and now, in the sixth edition by Mayne. Of these sections the Introduction covers interpretation of biochemical data and the last includes the details of test protocols.

The intervening 12 sections are arranged to correspond with the appropriate chapters in the sixth edition of *Clinical Chemistry in Diagnosis and Treatment*. Each of these 12 sections includes five case reports; in each group of five, the first cites laboratory data and is followed by questions which the student should attempt to answer before turning to the explanations given. The others omit the questions. Section 14 is a more orthodox data interpretation chapter, concentrating on biochemical results, and less on clinical information. All the cases are genuine and known to one of us.

We have included tables that detail causes of biochemical abnormalities; it is hoped that students will use these as an aid to revision while reading the appropriate case study and so reinforce their learning. We have also included some clinical photographs, X-rays and histological pictures.

Inevitably, some readers may not agree with all our interpretations, or even with the selection of tests. This should stimulate them to think for themselves - a desirable aim in itself.

Assays were performed in different hospital laboratories using methods with different reference ranges. This should remind readers always to interpret results using the reference range of the reporting laboratory, especially when patients move from one hospital to another.

We wish to thank Drs Mary Cafferkey, Michael Feher, Margaret Sinnet and Damien Griffin and Professors Joan Zilva and Rory O'Moore for their helpful and constructive comments, Professor Mary Leader for providing the histology slides and Ms Joanna Sheldon for preparing the protein electrophoreses. We are also grateful to the many students who have directly and indirectly helped us select the cases and clarify our own concepts and to Professor Joan Zilva and Dr Peter Pannall for allowing us to write this book as a companion to *Clinical Chemistry in Diagnosis and Treatment*. We are grateful to our colleagues for letting us publish data on patients under their care.

Finally we would like to thank the publishers and in particular Diane Leadbetter-Conway for their cooperation and support during the preparation of this book.

PDM
APD
March 1994

Interpretation of biochemical data

Investigations should be used:

- to confirm a diagnosis, arrived at following taking a history and performing a clinical examination;
- to assess the complications of a disorder;
- to monitor disease progression and treatment.

Biochemical investigations are, in themselves, rarely diagnostic of a clinical condition and should not be performed indiscriminately. At the time of requesting the test the clinician should have some idea as to what results to expect and of the significance of the results with regard to diagnosis and management. If they are very different, either he has made an error in his initial diagnosis or there has been an error in the handling of the specimen or in the reporting of the result.

A clinician should be aware of the many procedures involved in the collection, transport and analysis of blood specimens that may affect the result. However, some factors that affect the results of biochemical tests are discussed briefly.

1.1 APPROPRIATE METHODS OF COLLECTION OF BLOOD SAMPLES

There are many factors involved in the collection and transport of samples that may affect the analytical result and hence the conclusions reached when interpreting the result. Some of these are discussed.

Posture: because of the pooling of blood under gravity when an individual stands up, the results of some analytes, particularly proteins and substances bound to them, are slightly higher than those taken from an individual who has been lying down for some time. For this reason the results of some tests tend to be slightly higher in samples from out-patients and from General Practioners' surgeries compared with those from in-patients.

Oral medication: specimens should not be taken to measure an analyte or the effects of a drug just after an oral dose has been given. For example:

- blood should be taken for a drug assay at a standard time after the dose, the time being dependent on the half-life of the drug in blood;
- plasma potassium concentration may be low for a few hours after taking a potassium-losing diuretic because of the rapid clearance of potassium from the extracellular fluid.

Venesection: prolonged occlusion of a vein, using a tourniquet in order to make the vein 'stand out', may increase the plasma concentration of large molecules such as plasma proteins and those compounds which are partly bound to plasma proteins such as calcium. The effect of venesection on some plasma constituents is shown in Table 1.1.

Containers: most laboratories issue a list of the types of container required for different assays. Different containers, which are colour coded, contain different anticoagulants or preservatives. For example, for the measurement of plasma potassium concentration, blood is normally collected into a vial containing lithium

Table 1.1   Some effects of a tourniquet and of its release on plasma concentrations in a normal subject

| | Tourniquet on (minutes) | | | | | | Tourniquet off (minutes) | | |
	0	0.5	1	2	3	4	0.25	0.5	
Calcium	2.38	2.36	2.39	2.39	2.44	2.46	2.42	2.39	mmol/L
Total protein	81	81	83	83	88	89	86	84	g/L
Albumin	44	43	45	46	47	48	47	45	g/L
Glucose	4.2	4.3	4.2	3.6	3.6	3.6	4.1	4.1	mmol/L
Haemoglobin	13.7	13.8	14.0	14.3	14.6	15.0	14.6	14.0	g/dl
Haemotricrit	40.1	40.3	40.7	41.5	42.3	43.6	42.4	41.2	%

Table 1.2   Effect of delayed separation of plasma from a lithium heparinized blood specimen, maintained at room temperature.

| | Plasma separated after (hours) | | | | | | |
	0	2	4	6	10	24	
Potassium	4.1	4.2	3.9	4.3	4.8	6.4	mmol/L
Glucose	4.2	4.0	3.9	3.6	3.0	0.9	mmol/L

heparin, which prevents clotting and, therefore, the release of potassium from cells especially platelets, prior to separation of the plasma. Serum potassium concentrations are usually higher than those measured in plasma samples. If a blood glucose sample is not to be processed for at least two hours it should be collected into a vial containing fluoride. This inhibits glycolysis in the erythrocyte and consequently the plasma glucose concentration does not fall following phlebotomy.

Effects of delay in separating plasma from red cells: most clinicians are aware of the effect of haemolysis on the concentration of various constituents such as potassium in plasma, although this is variable. A similar effect may occur if there is a delay in separating plasma from the red blood cells (Table 1.2). The effect, to some extent, is dependent on the temperature at which the unseparated blood has been stored. This difference is not constant and may be marked. However, unlike a haemolysed sample, the plasma will not be red in colour and, therefore, the laboratory scientist will not immediately be alerted to the problem.

Having taken a blood sample with minimal stasis into the correct vial and transported it as quickly as possible to the laboratory the specimen must be centrifuged and the plasma either analysed immediately or stored for later analysis. Most common analytes are stable in plasma for a few days, but some require immediate freezing.

1.2   REFERENCE RANGES

A reference range, by definition, contains the results of 95 per cent of the values obtained from a 'reference' or defined population. Consequently, if the analyte has a normal or Gaussian distribution within the population 95 per cent of the results will fall within two standard deviations from the mean. Therefore, within any normal 'reference' population, five per cent of that population will have values outside the reference range, 2.5 per cent being below and 2.5 per cent above the range. This is an important concept because if a biochemical variable is outside the quoted reference range it does not necessarily mean that the result is abnormal. The more biochemical tests that are requested on a individual blood sample the greater the probability that at least one result will be outside a quoted range purely on the basis of statistics. It is important to remember that the further a result is from the population mean the more likely it is that there is a significant abnormality.

There are a number of factors which affect reference ranges. These include:

- *age*; for example plasma total alkaline phosphatase activity is about five times the upper adult reference limit in the neonate and falls to about twice the upper adult reference limit during childhood; it rises again during the adolescent growth spurt before falling to the adult range;
- *sex*; the plasma creatinine concentration, which is dependent on muscle bulk, tends to be lower in women than men; the sex steroids, oestradiol and testosterone, are obvious examples;
- *physiological variations such as the circadian rhythm*; some variables, such as plasma potassium or iron, fluctuate quite markedly from day to day. Others, such as cortisol, fluctuate in cyclical (circadian) rhythms throughout 24 hours, or in 28 day cycles, such as the sex steroids and gonadotrophins in women;
- *analytical precision*; all assays have an inherent imprecision. Therefore, rarely will exactly the same result be obtained if the sample is re-analysed. This analytical imprecision depends, to some extent, on the methodology and the instrumentation used.

Instrumentation and methodological differences may also affect the numerical value of biochemical tests. For example, although plasma enzyme activities are

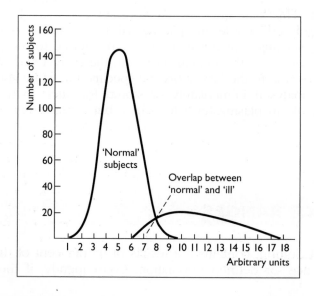

Fig. 1.1 Theoretical distribution of values for 'normal' and 'abnormal' subjects, showing overlap at the upper end of the reference range.

measured in 'international units' the plasma activities are dependent on many factors, such as the temperature at which the assay is performed, the concentration of substrate, the pH of the buffer or the presence of activators or inhibitors.

Because of these many factors, *biochemical results must be interpreted in relation to the appropriate reference range, quoted from the same laboratory.*

Having defined and established a reference range for the healthy population, it is important to know whether or by how much the values overlap those obtained from the diseased population. Ideally, there should be a complete separation between the two populations so that the variable has 100 per cent sensitivity (positive in the disease population) and 100 per cent specificity (negative in the healthy population) for the presence of disease. However, this rarely occurs and there is usually a considerable overlap between the values obtained from the ill population and from the healthy reference population, as shown in Fig. 1.1. The more the result of a biochemical test deviates from the mean, the greater is the probability that the test result is abnormal.

1.3   HOW TO INTERPRET BIOCHEMICAL DATA

Throughout this book, we use the term T_{CO_2} to denote the measured plasma bicarbonate concentration as it is this which is usually measured in venous plasma. T_{CO_2} is an estimate of the sum of the plasma concentrations of bicarbonate, carbonic acid and dissolved CO_2. At a pH of 7.4 the plasma T_{CO_2} concentration is approximately 1 mmol/L higher than the actual bicarbonate concentration.

Let us now consider the results of some biochemical tests obtained from a 28-year-old woman who was referred to the clinic for the investigation of hypertension.

Plasma			Reference range
Urea	5.7	mmol/L	2.5–7.0
Sodium	143	mmol/L	135–145
Potassium	2.8	mmol/L	3.5–4.8
T_{CO_2} (bicarbonate)	35	mmol/L	22–32

The most striking feature of the biochemical results is the relatively low plasma potassium concentration at 2.8 mmol/L with respect to the quoted reference range. There are many causes of a low plasma potassium concentration. However, the differential diagnosis can be greatly reduced by considering whether the low plasma potassium concentration is associated with a metabolic alkalosis (a raised plasma T_{CO_2} concentration) or with a metabolic acidosis (a low plasma T_{CO_2} concentration). In this patient the plasma T_{CO_2} (bicarbonate) concentration is raised at 35 mmol/L, the findings being compatible with hypokalaemic alkalosis. The differential diagnosis can be reduced further by considering the plasma sodium concentration, which reflects the relative proportion of water to sodium within the extracellular compartment rather than the total amount of sodium present. Although the sodium concentration is within the reference range it is at the upper end.

The relatively high plasma sodium concentration, taken together with the hypokalaemic alkalosis, is suggestive of excess mineralocorticoid activity. Enhanced sodium reabsorption in the distal renal tubules in exchange for potassium with iso-osmotic water reabsorption, is a probable cause of the hypertension. In contrast, if the plasma sodium concentration had been below 140 mmol/L, for example 136 mmol/L, these biochemical changes might have been caused by a potassium-losing diuretic such as a thiazide, used in the treatment of hypertension.

This example is used to illustrate the amount of information that can be derived from a minimal set of biochemical data. Let us consider another set of biochemical results obtained from a 54-year-old man who has a long history of chronic renal failure.

Plasma			Reference range
Creatinine	512	µmol/L	60–120
Sodium	134	mmol/L	135–145
Potassium	5.5	mmol/L	3.5–4.8
T_{CO_2} (bicarbonate)	16	mmol/L	22–32
Calcium	1.89	mmol/L	2.15–2.55
Albumin	31	g/L	32–45
Phosphate	1.56	mmol/L	0.60–1.40
Urate	0.62	mmol/L	0.17–0.44

The results of all these tests are outside the reference ranges. However, the reference ranges quoted are those for a healthy reference population and not for a population of patients with renal dysfunction. Having established that the patient has renal glomerular dysfunction, because of the significantly raised plasma creatinine concentration, the rest of the tests should be interpreted against the ranges found in patients with renal glomerular dysfunction. In this patient, there is a hyperkalaemic acidosis, a low plasma calcium concentration, despite a relatively normal plasma albumin associated with a raised plasma phosphate concentration and a raised plasma urate concentration. However, these findings are due to the metabolic complications that frequently occur in patients with renal glomerular dysfunction. The plasma sodium concentration tends to be slightly lower than that in healthy individuals; consequently a plasma sodium of 134 mmol/L might be considered appropriate for a patient with a plasma creatinine concentration of 512 µmol/L. However, if the plasma sodium concentration had been 143 mmol/L, this would have suggested that the patient had relative water depletion, which might have contributed to a reduction in the glomerular filtration rate (GFR) and exacerbated the renal glomerular impairment. Cautious hypotonic fluid replacement might increase the GFR with a reduction in the plasma creatinine concentration and a possible improvement in some of the metabolic complications such as the hyperkalaemic acidosis and the hyperuricaemia.

This example illustrates that having confirmed or made an initial diagnosis, which has affected the results of one or more biochemical variables, the results of the other biochemical tests should be interpreted in relation to those expected in that condition rather than those found in the healthy reference population. Furthermore, it is more likely that there is a single cause to explain all the biochemical changes rather than many different primary disorders.

These are just two examples of how students and clinicians should interpret biochemical data, maximizing the information obtained. We hope that these basic principles will be applied when interpreting the data presented in this book. However, it is also important that the reader has a basic knowledge of the underlying physiology and pathophysiology associated with the conditions discussed in the clinical cases.

Renal dysfunction

CASE 2.1

A 48-year-old woman presented with a six-month history of thirst and nocturia. She had also noticed mild ankle swelling on several occasions and had felt generally unwell. Her past medical history was unremarkable apart from hypertension, which had been detected five years previously; this had been moderately well controlled with nifedipine. Examination confirmed mild ankle oedema. Her blood pressure was 160/115 mmHg. There were no other clinical findings. Urine dipstick testing demonstrated haematuria and proteinuria. The following biochemical investigations were carried out.

Plasma			Reference range
Creatinine	261	µmol/L	60–100
Urea	18.4	mmol/L	3.0–7.0
Sodium	139	mmol/L	133–143
Potassium	5.4	mmol/L	3.6–4.6
T_{CO_2} (bicarbonate)	19	mmol/L	24–32

QUESTIONS:

1. What is your interpretation of these results?
2. How would you investigate the cause of these abnormalities?
3. What other biochemical abnormalities would you expect to find, and how would you explain them?

1. *Interpretation of the results.* The plasma creatinine and urea concentrations are increased proportionally above their respective reference ranges. This almost certainly indicates significant renal glomerular dysfunction, since these two substances are filtered by the glomeruli and are only minimally affected by tubular handling. It is unlikely to be due to reduced renal blood flow, as in this situation the rise in plasma urea concentration is disproportionately increased compared to that of creatinine as a low renal tubular flow allows time for increased reabsorption of urea from the collecting ducts and relatively greater creatinine secretion from the normal tubules.

There is also mild hyperkalaemia and a low plasma T_{CO_2} concentration, consistent with a metabolic acidosis. These are both features of chronic renal failure. The plasma potassium concentration is frequently only slightly raised since the ability to secrete potassium ions at the distal tubule is preserved until end-stage oliguric renal damage develops; there may also be a degree of proximal tubular dysfunction that reduces potassium reabsorption. In contrast, in acute renal failure hyperkalaemia may be severe and life-threatening. The plasma potassium concentration is also related to the hydrogen ion concentration, or pH, in the extracellular fluid. Metabolic acidosis due to a reduced renal capacity to excrete hydrogen ions is a feature of both glomerular and tubular failure. Glomerular dysfunction reduces filtration of buffer anions and, therefore, the availability of sodium ions for exchange with hydrogen.

2. *Further investigations to identify the cause of renal glomerular dysfunction.* An intravenous pyelogram (IVP) was performed and showed bilateral polycystic kidneys. This finding was confirmed by renal ultrasound. Polycystic disease of the kidneys may be inherited as an autosomal recessive disorder of varying severity. On further enquiry it transpired that this patient's mother also had chronic renal failure and a maternal aunt probably died with the same condition.

3. *The following additional biochemical investigations were performed:*

Plasma			Reference range
Calcium	2.19	mmol/L	2.25–2.70
Albumin	42	g/L	35–55
Phosphate	1.50	mmol/L	0.85–1.40
Alkaline phosphatase (ALP)	76	U/L	21–92

Mild hypocalcaemia is a common finding in chronic renal failure due to reduced activity of the enzyme 1α-hydroxylase which is present in the cells lining the proximal tubules; it converts 25-hydroxyvitamin D to the active metabolite 1,25-dihydroxyvitamin D. In chronic renal failure there is reduced enzyme activity because of:

- damage to the proximal tubular cells;
- inhibition by the raised phosphate concentration.

Low plasma 1,25-dihydroxyvitamin D concentrations result in reduced calcium absorption from the intestine. However, the fall in plasma calcium concentration is limited by increased mobilization of calcium from bone and some increase in renal tubular reabsorption of calcium due to a rise in parathyroid hormone secretion (secondary hyperparathyroidism). This effect may be reduced when the concentration of 1,25-dihydroxyvitamin D is low.

Hyperphosphataemia is almost invariably present in chronic renal dysfunction due to a reduction in the rate of filtration. It can be controlled by dietary restriction and the use of oral phosphate binders, such as aluminium hydroxide. It is important that the plasma phosphate concentration is reduced before any attempt is made to treat hypocalcaemia, since simultanously high calcium and phosphate concentrations may result in precipitation of calcium phosphate in extra-osseous tissues such as the kidneys, causing further deterioration in renal function.

Plasma alkaline phosphatase activity was normal. As renal failure progresses, renal osteodystrophy develops because of 1,25-dihydroxyvitamin D deficiency and secondary hyperparathyroidism. The resulting osteoblastic proliferation leads to a rise in plasma alkaline phosphatase activity.

The diagnosis of chronic renal failure can be made on the basis of the clinical and biochemical findings and radiological investigations. The elevated plasma creatinine concentration indicates that the glomerular filtration rate (GFR) is low. The exact relation between plasma creatinine concentration and GFR for an individual patient depends on the rate of creatinine production, which is a function of muscle mass. It may *occasionally* be useful to assess the glomerular filtration rate more directly by measuring the creatinine clearance.

The following results were obtained:

Plasma creatinine	264	μmol/L
Urinary creatinine	3.76	mmol/L
24-hour urine volume	2.73	L

$$\text{Creatinine clearance} = \frac{\text{urinary [creatinine]} \times \text{urinary volume}}{\text{plasma [creatinine]} \times \text{time}}$$
$$= 27 \ \text{ml/minute}$$

Figure 2.1 shows the relation between creatinine clearance and the plasma creatinine concentration. When there is early glomerular dysfunction, a significant

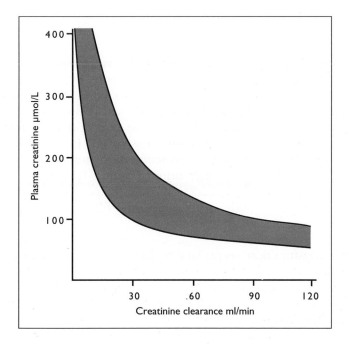

Fig. 2.1 The relation between creatinine clearance and the plasma creatinine concentration showing the approximate 95 per cent confidence intervals

fall in creatinine clearance results in a relatively small increase in plasma creatinine concentration. However, there is a large increase in the plasma creatinine concentration associated with a small fall in the creatinine clearance when there is severe glomerular dysfunction. Because of the problems associated with the collection of an accurately timed urine sample, the measurement of the plasma creatinine concentration is preferable to serial measurements of creatinine clearance in monitoring glomerular function in patients with severe chronic renal glomerular dysfunction.

Progress: she was started on additional antihypertensive medication, in order to reduce the rate of deterioration of renal function. This was monitored by measuring changes in the plasma creatinine concentration.

CASE 2.2   ACUTE RENAL DYSFUNCTION

A 75-year-old woman was admitted to hospital with a three-week history of nausea and vomiting. She was clinically volume depleted and oliguric. Intravenous rehydration was commenced with isotonic saline and five per cent dextrose; within seven days urine output increased to more than three litres per day. The results of serial biochemical changes were as follows:

Plasma	Day following admission						Reference range
	1	2	3	5	7		
Creatinine	868	745	533	438	321	μmol/L	60–100
Urea	44.2	38.3	28.0	23.7	15.6	mmol/L	3.0–7.0
Sodium	131	135	136	140	141	mmol/L	133–143
Potassium	6.9	5.5	4.2	3.4	2.9	mmol/L	3.6–4.6
$T\text{CO}_2$ (bicarbonate)	8	10	9	9	10	mmol/L	24–32

Comment: plasma urea and creatinine concentrations were high on admission, indicating renal glomerular dysfunction. There was severe hyperkalaemia and a low plasma $T\text{CO}_2$ concentration, consistent with a metabolic acidosis. These abnormalities are consistent with acute renal dysfunction.

During rehydration the plasma urea and creatinine concentrations fell as renal perfusion and GFR increased. There was a progressive fall in the plasma potassium concentration but the plasma $T\text{CO}_2$ concentration remained low. These findings, together with the abnormally high urinary output (polyuria), may occur during recovery from acute renal dysfunction due to persistent renal tubular dysfunction or to the transient osmotic effect due to the filtration of large amounts of previously retained urea.

Diagnosis: acute renal dysfunction secondary to intravascular volume depletion.

CASE 2.3   RENAL CIRCULATORY INSUFFICIENCY ('PRERENAL URAEMIA')

A 67-year-old man had a hemicolectomy for carcinoma of the colon. By the sixth postoperative day daily urinary output had decreased to 300 ml.

Plasma	On admission	Six days postoperative	Reference range
Urea	8.0	24.6 mmol/L	2.5–8.0
Sodium	138	141 mmol/L	135–145
Potassium	4.2	5.1 mmol/L	3.5–4.8
T_{CO_2} (bicarbonate)	23	19 mmol/L	22–32

Urine			
Sodium		9 mmol/L	

Comment: by the sixth day there had been a rise in the plasma urea concentration, associated with a hyperkalaemic metabolic acidosis and a slight increase in plasma sodium concentration. These biochemical changes could have been caused either by renal circulatory insufficiency ('prerenal uraemia') due to intravascular volume depletion, or by renal glomerular dysfunction. Some causes of a raised plasma urea concentration are shown in Table 2.1.

The measurement of plasma creatinine and urinary sodium concentrations may distinguish between a prerenal uraemia and established renal damage. In prerenal uraemia due to intravascular volume depletion, the plasma urea concentration usually rises faster than that of creatinine and the urinary sodium concentration is low. Intravascular volume depletion stimulates renin and aldosterone secretion resulting in increased reabsorption of sodium from the distal renal tubular lumina, thus reducing the urinary sodium concentration. Consequently a urinary sodium concentration of less than about 20 mmol/L in the presence of clinical evidence of hypovolaemia is suggestive of a prerenal

Table 2.1   Some causes of a raised plasma urea concentration

Prerenal uraemia
    decreased glomerular filtration rate
        intravascular volume depletion
    increased protein breakdown (catabolism)
        infection
        steroids
        large gastrointestinal haemorrhage
        major trauma
Intrinsic renal disease
    glomerular dysfunction
Postrenal uraemia
    renal tract obstruction
    ureteric strictures
    carcinoma of the bladder
    enlargement of the prostate

cause of uraemia with adequately functioning tubules rather than intrinsic renal glomerular dysfunction.

Diagnosis: renal circulatory insufficiency ('prerenal uraemia') secondary to intravascular volume depletion.

CASE 2.4   CHRONIC RENAL DYSFUNCTION

A 36-year-old woman was referred to the out-patient clinic with a six-month history of progressive lassitude and shortness of breath. There was no previous significant past medical history. On examination she was pale with a rather sallow appearance, a blood pressure of 170/100 mmHg and a regular pulse of 92 per minute.

Plasma			Reference range
Urea	42.2	mmol/L	2.5–7.0
Sodium	132	mmol/L	135–145
Potassium	6.7	mmol/L	3.5–4.8
T_{CO_2} (bicarbonate)	12	mmol/L	22–32
Calcium	1.78	mmol/L	2.15–2.55
Albumin	32	g/L	32–45
Phosphate	3.8	mmol/L	0.60–1.40
Blood			
Haemoglobin	6.2	g/dl	11.5–16.5

Comment: the raised plasma urea concentration and the hyperkalaemic metabolic acidosis (low plasma T_{CO_2} concentration) are compatible with glomerular dysfunction.

The clinical history and biochemical findings are consistent with chronic intrinsic renal damage. Glomerular damage results in retention of urea and creatinine, the concentrations of which rise in plasma. Reduced filtration of sodium ions results in reduced availability of sodium for reabsorption in exchange for potassium in the distal renal tubules and so hyperkalaemia develops. Impaired excretion of hydrogen ions and generation of bicarbonate ions contributes to the metabolic acidosis (reduction in plasma T_{CO_2} concentration) and the rise in plasma potassium concentration. The plasma sodium concentration tends to be at the lower end of the reference range in patients with renal glomerular dysfunction who are normovolaemic. A high normal or high plasma sodium concentration is suggestive of water depletion and this may contribute to the rise in plasma urea concentration.

As renal glomerular damage progresses, associated with a reduced glomerular filtration rate, the plasma phosphate concentration rises and plasma calcium concentration falls. The fall in plasma calcium concentration may be caused by:

- a decrease in the plasma concentration of the active vitamin D metabolite (1,25-dihydroxyvitamin D). The hydroxylation of 25-hydroxyvitamin D takes place in the proximal renal tubular cells; the enzyme is inhibited by a high phosphate concentration;

- a reduction in the mass of functioning renal tissue and, therefore, in the ability of the kidneys to synthesize the active vitamin D metabolite;
- a rise in the plasma phosphate concentration, which alters the solubility equilibrium of calcium and phosphate and which may cause its precipitation.

Intrinsic renal damage is associated with a fall in erythropoietin synthesis and consequently a reduction in erythropoiesis and low blood haemoglobin concentration.

Diagnosis: chronic renal dysfunction.

CASE 2.5   NEPHROTIC SYNDROME

A 10-year-old boy was referred because of a recent onset of marked swelling of his ankles. He had never been in hospital before and was on no medication. On examination there was pitting oedema of both legs up to his mid-calf. He was normotensive. Protein, but no blood, was found in the urine. A provisional diagnosis of nephrotic syndrome was made and the following investigations were performed.

Plasma			**Reference range**
Creatinine	78	μmol/L	50–110
Urea	4.5	mmol/L	2.5–6.5
Sodium	137	mmol/L	135–145
Potassium	4.0	mmol/L	3.5–4.8
Total protein	44	g/L	60–80
Albumin	18	g/L	32–45
Calcium	2.01	mmol/L	2.15–2.65
Phosphate	1.41	mmol/L	1.20–1.95
ALP	464	U/L	170–800

Urine		
Total protein	17.6 g/24 hours	<0.15

Interpretation: these biochemical results confirm the clinical diagnosis. The plasma total protein and albumin concentrations are significantly reduced and there is proteinuria in excess of 5.0 g/24 hours. The plasma calcium concentration is low but, when corrected for the low plasma albumin concentration, it is probably normal, indicating that it is low because of the low plasma albumin concentration to which 50 per cent of the calcium is bound in plasma. The plasma free-ionized fraction is almost certainly within the reference range. The plasma alkaline phosphatase (ALP) activity was normal for his age.

Comment: causes of the nephrotic syndrome are shown in Table 2.2. During childhood the majority of cases of nephrotic syndrome are caused by minimal change glomerulonephritis, a condition which usually responds to steroid treatment. The severity of the glomerular lesion can be determined by measuring the differential protein clearance or selectivity index, the proportion or ratio of low molecular-weight proteins (albumin or transferrin) in relation to high molecular-weight

Table 2.2   Some causes of the nephrotic syndrome

Primary glomerular disease (rarer)
  minimal change glomerulonephritis
  membraneous glomerulonephritis
  diffuse glomerulonephritis
  mesangiocapillary glomerulonephritis
Glomerular damage secondary to other diseases
  diabetes mellitus
  amyloidosis
  Henoch–Schönlein purpura
  systemic lupus erythematosus
  infections
    malaria
    bacterial endocarditis
  drugs
    gold, penicillamine

proteins (IgG) lost through the glomeruli. The measurement of a ratio, rather than an excretion rate, obviates the need to collect a timed urine sample. The selectivity index is probably better in predicting the response to treatment than examination of a needle biopsy specimen of the kidney; it is a noninvasive technique, with minimal morbidity.

The serum albumin concentration was estimated again and found to be 17.0 g/L. The following additional investigations were performed:

Serum
Albumin   17.0 g/L
IgG   3.6 g/L

Urine
Albumin   15.8 g/L
IgG   0.16 g/L

Selectivity index

$$\frac{serum[Alb] \times urine[IgG]}{urine[Alb] \times serum[IgG]} = 0.048$$

Interpretation:
<0.16   selective
 0.16–0.30   moderately selective
>0.30   nonselective

Serum and urine protein electrophoresis
Reduced serum albumin with increased α_2-globulin, albuminuria (Fig. 2.2)

Comment: this boy had a selective proteinuria with typical changes on protein electrophoresis. The increased serum α_2-globulin is caused by the increased synthesis of high molecular-weight (720 000) α_2-macroglobulin. Histological examination of a renal biopsy specimen confirmed the cause as minimal change glomerulonephritis. He responded to treatment with steroids.

Diagnosis: nephrotic syndrome with highly selective proteinuria, caused by minimal change glomerulonephritis.

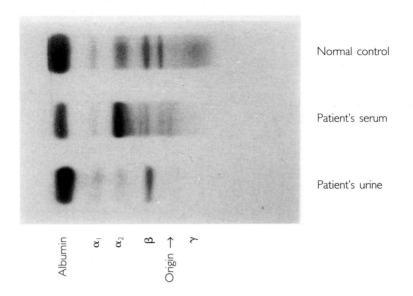

Normal control

Patient's serum

Patient's urine

Albumin α_1 α_2 β Origin → γ

Fig. 2.2   Serum and urine protein electrophoresis from a child with nephrotic syndrome, showing a reduced serum albumin and an increased α_2-globulin concentration and a selective glomerular proteinuria with excretion of low molecular-weight proteins, predominantly albumin and transferrin

Disorders of sodium and water homeostasis

3

CASE 3.1

A 67-year-old woman with bronchiectasis was admitted to hospital with a two-week history of a productive cough with green sputum. Over the previous week, she had become confused and disorientated.

On examination her mental responses were slow. Her blood pressure was 150/80 mmHg. She was neither clinically volume depleted nor oedematous. There were bilateral widespread coarse crepitations in the lungs, but no other significant clinical abnormalities.

The following results were obtained:

Plasma			Reference range
Creatinine	64	μmol/L	55–110
Urea	2.4	mmol/L	2.5–7.0
Sodium	122	mmol/L	135–145
Potassium	4.0	mmol/L	3.5–4.8

QUESTIONS:

1. Could these results explain the clinical presentation?
2. What are the possible causes of the hyponatraemia?
3. How would you investigate this patient further?

1. *Clinical interpretation of biochemical results.* Confusion, ataxia, dysarthria, and apathy can all be attributed to cerebral cellular overhydration secondary to hyponatraemia. However, a plasma sodium concentration of 122 mmol/L is not usually associated with many, or even any, symptoms. The risk of developing clinical symptoms and signs is related more to the rate of fall of the plasma sodium concentration than to its absolute level.

2. *Causes of hyponatraemia* are shown in Table 3.1. In this case the following causes should be considered.

Hyponatraemia with normal plasma osmolality

- *Pseudohyponatraemia*, caused by analytical interference by high concentrations of lipid or protein. Because the extracellular fluid (ECF) osmolality is normal, there is no movement of water into the intracellular compartment.

Hyponatraemia with increased plasma osmolality

- *Appropriate hyponatraemia* due, for example, to hyperglycaemia. This is a dilutional hyponatraemia due to the movement of water out of the cellular compartment along the osmotic gradient generated by extracellular hyperglycaemia. A similar situation can occur in patients with uraemia despite a high proportion of functioning glomeruli.

Hyponatraemia with decreased plasma osmolality

- *Diuretics* (thiazide and loop diuretics). These diuretics inhibit renal tubular sodium, and therefore water, reabsorption; as a consequence ECF volume is reduced which, by stimulating antidiuretic hormone (ADH) release, increases water reabsorption and causes a dilutional hyponatraemia.
- *Adrenocortical hypofunction*, due to:
    primary adrenal hypofunction in which hyponatraemia is a consequence of acute ECF volume depletion, ADH release and subsequent water reabsorption;
    secondary adrenal hypofunction, in which aldosterone secretion is normal,

Table 3.1   Some causes of hyponatraemia

Without plasma hypo-osmolality
   with iso-osmolality
     pseudohyponatraemia
   with hyperosmolality
     diabetes mellitus
     renal failure
With hypo-osmolality
   with extracellular fluid (ECF) volume depletion
     diuretics
     salt-losing nephritis
     adrenal insufficiency
   with undetectable increased ECF volume
     syndrome of inappropriate ADH secretion (SIADH)
   with ECF volume expansion
     cardiac failure
     renal failure
     nephrotic syndrome

---

but cortisol deficiency results in dilutional hyponatraemia as a consequence of reduced renal water excretion.

- *Severe primary hypothyroidism*, associated with a dilutional hyponatraemia due to increased ADH secretion.
- *Inappropriate ADH secretion.* This diagnosis should be considered after the exclusion of other causes of hyponatraemia (Table 3.2). Hyponatraemia may occur in any severe illness. However, in this syndrome, ADH is secreted inappropriately, either from ectopic sites or from the posterior pituitary gland, in conditions when its secretion would be normally suppressed by the low plasma osmolality (Table 3.3).

3. *The following additional investigations were requested:*

- *plasma osmolality and calculated osmolarity.* Approximate plasma osmolarity can be calculated using the formula:

$$\text{Osmolarity} = 2 \times ([Na^+] + [K^+]) + [\text{urea}] + [\text{glucose}]$$

Table 3.2 Some causes of hyponatraemia that must be excluded before a diagnosis of inappropriate ADH secretion is made

---

Any severe illness that may cause SIADH
Hypovolaemia
Oedematous disorders
    cardiac failure
    cirrhosis of the liver
Endocrine disorders
    hypothyroidism
    adrenal insufficiency
Renal failure
Drugs that impair water excretion
    diuretics
    psychotropic drugs

Table 3.3 Some disorders which may be associated with inappropriate ADH secretion. However, any, even mild, illness may cause water retention

---

Neoplasia
    bronchogenic carcinoma
    islet cell tumours of the pancreas
    lymphoma
Pulmonary disorders
    pneumonia
    tuberculosis
    pneumothorax
    positive pressure ventilation
Neurological disorders
    encephalitis
    meningitis
    head injury
    acute intermittent porphyria

If the measured plasma osmolality is normal but the calculated osmolarity is low and if there is no unmeasured solute such as ethanol, pseudohyponatraemia due to an increased concentration of either protein or lipid is the most likely cause;

- *urinary osmolality.* In ADH-induced hyponatraemia, plasma osmolality is low and urinary osmolality inappropriately high. If the kidneys are functioning normally the following diagnostic criteria must be satisfied before a diagnosis of hyponatraemia due to inappropriate ADH secretion is made:

    absence of signs of volume depletion or oedema;

    normal renal, adrenal and thyroid function;

    hyponatraemia and a low plasma osmolality;

    inappropriately high urinary osmolality when compared with that of plasma;

    high urinary sodium concentration.

- *urinary sodium concentration.* If it is less than about 10 mmol/L, ECF volume depletion due to nonrenal fluid loss must be considered. It is normally greater than about 20 mmol/L in diuretic-induced hyponatraemia, adrenocortical insufficiency, or conditions associated with inappropriate ADH secretion;

- *plasma glucose concentration.* The plasma sodium concentration may fall, due to dilution, as the plasma glucose concentration, and therefore osmolality, rises provided that adequate fluid intake is maintained;

- *thyroid function tests* (plasma TSH and free or total T_4 concentrations). It is necessary to exclude hypothyroidism because she presented with some clinical features of hypothyroidism.

Later results:

Plasma			Reference range
Creatinine	64	μmol/L	55–110
Urea	2.4	mmol/L	2.5–7.0
Sodium	122	mmol/L	135–145
Potassium	4.0	mmol/L	3.5–4.8
Glucose	3.9	mmol/L	3.6–6.5
Osmolality	261	mmol/kg	275–295
Calculated osmolarity	258	mmol/L	
Total T_4	64	nmol/L	60–140
TSH	1.5	mU/L	0.5–2.9

Urine		
Osmolality	560	mmol/kg
Sodium	55	mmol/L

Comment: these results satisfy the diagnostic criteria for hyponatraemia due to inappropriate ADH secretion. The urinary osmolality is inappropriately high for the relatively low plasma osmolality. Low–normal plasma urea and creatinine concentrations indicate that the patient is well hydrated and has normal renal function. The urinary sodium concentration is appropriate for a normovolaemic subject. Adrenocortical hypofunction was considered unlikely because of her clinical presentation and the normal plasma potassium concentration.

Progress: Haemophilus influenzae was cultured from her sputum and appropriate antibiotics were given. Fluid intake was restricted to about 800 ml per day. After two days she developed a negative water balance and the plasma sodium concentration and osmolality rose to within the reference range after six days; her mental function improved significantly.

Plasma	On admission	Day 6	
Creatinine	64	70	µmol/L
Urea	2.4	3.4	mmol/L
Sodium	122	137	mmol/L
Potassium	4.0	4.5	mmol/L
Osmolality	261	292	mmol/kg

Comment: the management of symptomatic hyponatraemia is to treat the underlying disorder and to restrict water intake. A water diuresis can be promoted by using a drug, demeclocycline, which causes a nephrogenic diabetes insipidus by inhibiting the action of ADH on the renal collecting ducts. It is rarely necessary, and often dangerous, to infuse a hypertonic saline solution, with or without diuretics.

Diagnosis: inappropriate ADH secretion due to bronchopneumonia in a patient with bronchiectasis.

CASE 3.2   CONGESTIVE CARDIAC FAILURE (CCF)

An 85-year-old woman was admitted to hospital with congestive cardiac failure having developed progressive dyspnoea on exertion, nocturia and marked ankle oedema. Clinical examination was consistent with the history; she was in rapid atrial fibrillation. She was started on digoxin and the diuretic Frumil, which contains a potassium-sparing diuretic amiloride and the loop diuretic frusemide. The results of the laboratory investigations over the first four days are shown below.

Plasma	Day 1	Day 2	Day 4		Reference range
Creatinine	115	133	236	µmol/L	55–110
Urea	9.4	15.7	24.3	mmol/L	2.5–7.0
Sodium	129	128	121	mmol/L	135–145
Potassium	4.4	4.8	6.2	mmol/L	3.5–4.8

Comment: during the first four days there was a progressive deterioration in renal function; plasma urea concentration rose to 24.3 mmol/L and creatinine to 236 µmol/L. This was associated with a fall in plasma sodium from 129 to 121 mmol/L and a significant increase in plasma potassium concentration from 4.4 to 6.2 mmol/L. The differential diagnosis to explain the biochemical changes includes:

- *volume depletion with a prerenal uraemia.* Volume depletion, caused by an excessive diuresis, could have caused a reduction in the glomerular filtration rate with a rise in plasma urea concentration. Volume depletion stimulates antidiuretic hormone secretion, resulting in increased water reabsorption from the renal collecting ducts, causing a dilutional hyponatraemia. The plasma potassium concentration of 6.2 mmol/L is inappropriately high for the rise in plasma urea and could have been caused by the potassium-sparing diuretic, amiloride;

- *acute adrenal insufficiency.* A short tetracosactrin test was performed to exclude adrenal insufficiency (p. **180**).

Short tetracosactrin ('Synacthen') test

Time	0	30	60	minutes
Cortisol	475	911	1019	nmol/L

Interpretation: normal response. Following 250 μg 'Synacthen' intramuscularly, the plasma cortisol concentration should increase by at least 200 nmol/L to a concentration greater than 550 nmol/L by 30 minutes.

Management: Frumil was stopped and she was started on frusemide only; the subsequent biochemical changes are shown below.

Plasma	Day 1	Day 2	Day 4	Day 9	
Creatinine	115	133	236	91	μmol/L
Urea	9.4	15.7	24.3	6.1	mmol/L
Sodium	129	128	121	126	mmol/L
Potassium	4.4	4.8	6.2	4.1	mmol/L
			Frumil stopped		

Comment: five days after changing the diuretic, there had been a significant clinical improvement with a marked reduction in the ankle oedema associated with a loss in weight. The plasma urea and creatinine concentrations were now normal and the plasma potassium had fallen to 4.1 mmol/L.

Diagnosis: volume depletion secondary to excessive and inappropriate type of diuretic therapy resulting in a prerenal uraemia and a dilutional hyponatraemia.

CASE 3.3   CRANIAL DIABETES INSIPIDUS

A 23-year-old nurse was referred to the out-patient clinic with a six-month history of progressive polyuria with nocturia. She claimed that she drank 3.5–4 litres fluid per day. The following biochemical investigations were performed.

Plasma			Reference range
Urea	5.3	mmol/L	2.5–7.0
Sodium	144	mmol/L	135–145
Potassium	3.6	mmol/L	3.5–4.8
T_{CO_2} (bicarbonate)	28	mmol/L	22–32

Urine		
Osmolality	74	mmol/kg
Sodium	29	mmol/L

Table 3.4   Some causes of polyuria

Impaired ADH production
    cranial diabetes insipidus
        congenital
        acquired
Impaired renal tubular response to ADH
    nephrogenic diabetes insipidus
        congenital
        acquired
            interstitial nephritis
            renal medullary cystic disease
            hypercalcaemia or hypokalaemia
            drugs
                lithium carbonate
                demeclocycline
    chronic renal failure
Osmotic diuresis
    glycosuria (diabetes mellitus)
    uraemia
    mannitol treatment
Other drugs
    diuretics

Interpretation: although these results were apparently normal the plasma sodium concentration was at the upper end of the reference range; the urine osmolality was low.

Comment: some causes of polyuria are shown in Table 3.4. In this patient, it was unlikely that the polyuria was caused by an osmotic diuresis due to diabetes mellitus as the urinary osmolality was only 74 mmol/kg; the presence of glycosuria should have been tested for. If diuretic abuse and hypercalcaemia could be excluded the most likely diagnosis was diabetes insipidus.

Because the results of the initial investigations were equivocal, a water deprivation test was performed to investigate the possibility of diabetes insipidus (p. **172**). The patient was deprived of water from 23.00 hours the previous evening. Plasma and urine samples were collected from 07.00 hours the following day. The results are shown below:

Water deprivation test

Time	Osmolality mmol/kg	
(hours)	plasma	urine
07.00	292	72
08.00	295	78
09.00	296	76

Intramuscular injection of synthetic ADH (DDAVP)

11.00	294	814
12.00	295	998

Interpretation of water deprivation test: although there was a slight increase in the plasma osmolality from 292 to 296 mmol/kg there was no significant change in the urinary osmolality, which remained inappropriately low. This confirmed the diagnosis of some form of diabetes insipidus. Following an intramuscular injection of DDAVP (1-**D**eamino-8-**D**-**A**rginine **V**aso**p**ressin; desmopressin acetate) the urinary osmolality exceeded 850 mmol/kg demonstrating that urine concentration could occur in the presence of an ADH analogue. A diagnosis of cranial diabetes insipidus was made; additional investigations were performed to identify the cause of the disorder.

Diagnosis: cranial diabetes insipidus.

CASE 3.4 WATER DEPLETION

An 84-year-old man, with senile dementia, had a repair of an inguinal hernia. He subsequently developed a wound infection, with pyrexia, and a sacral bedsore. He became increasingly withdrawn over the next two weeks. On day 14, his urine output was recorded as being less than 300 ml per 24 hour. Plasma urea and electrolyte concentrations were measured:

Plasma	On admission	Day 14		Reference range
Urea	8.5	21.8	mmol/L	3.0–7.0
Sodium	141	161	mmol/L	133–143
Potassium	4.3	3.7	mmol/L	3.6–4.6
Chloride	101	120	mmol/L	95–105
T_{CO_2} (bicarbonate)	28	19	mmol/L	24–32

Interpretation: the slightly high plasma urea concentration on admission probably indicates mild renal glomerular dysfunction, which is a common finding in the elderly. By day 14 he had developed hypernatraemia with a significant rise in the plasma urea concentration. There was also a fall in his plasma T_{CO_2} concentration, consistent with a mild metabolic acidosis.

Comment: some causes of hypernatraemia are shown in Table 3.5. Hypernatraemia is usually caused by intravascular water depletion due either to inadequate intake or increased loss. Excessive sodium intake and inappropriate sodium retention by the kidneys, without adequate water, may also cause hypernatraemia. However, this only occurs if there is failure to respond appropriately to the increase in plasma osmolality, with stimulation of thirst and hence water intake, and ADH mediated reabsorption of water from the renal collecting ducts.

Sodium is by far the most abundant extracellular cation; hypernatraemia *always* indicates extracellular hyperosmolality. The increased osmotic gradient across cell membranes produces intracellular dehydration. Once hypernatraemia developed, the increased extracellular osmolality and consequent shift of water out of the intracellular compartment contributed to the deterioration in the patient's mental state.

The most likely cause for the hypernatraemia was water depletion due to failure to drink. Normally, a small increase in extracellular osmolality stimulates thirst and the water deficit is corrected. This patient's thirst response was probably impaired due to senile dementia and his mental confusion was probably increased by the

Table 3.5 Some causes of hypernatraemia

Inadequate water intake
   lack of water
   inability to drink
   impaired thirst response
Sodium-free water loss
   increased renal loss
      diabetes insipidus
      osmotic diuresis
   nonrenal loss of hypotonic fluid
      excessive sweating
      burns
      hyperventilation
      diarrhoea
      vomiting
Excessive sodium intake
   diet
   drugs (especially antibiotics)
Excessive sodium retention
   hyperaldosteronism

stress of admission to hospital and the development of a wound infection. The fluid balance chart confirmed a very low oral fluid intake. An additional cause of the hypernatraemia was increased sweating due to pyrexia; the concentration of sodium in sweat is much lower than in plasma.

Diagnosis: relative water depletion due to reduced water intake and increased water loss caused by pyrexia and sweating.

CASE 3.5 POSTOPERATIVE HYPONATRAEMIA

An 83-year-old noninsulin-dependent diabetic woman underwent an anterior resection of the bowel for carcinoma of the rectum. During the postoperative period she required insulin to control the blood glucose concentration. On the fourth postoperative day, she became confused and agitated. Blood was taken for biochemical analysis and the results were reviewed with those taken preoperatively and immediately postoperatively.

Plasma	Preoperative	Day 1	Day 4		Reference range
Urea	6.2	4.0	1.5	mmol/L	3.0–7.0
Sodium	135	129	124	mmol/L	135–145
Potassium	4.5	3.9	3.5	mmol/L	3.5–4.8
TCO_2 (bicarbonate)	32	30	31	mmol/L	24–32
Glucose	11.5	4.6	7.4	mmol/L	2.2–5.5

Interpretation: the results obtained preoperatively were essentially normal apart from a slightly increased plasma glucose concentration. During the four days after surgery, there was a progressive fall in the plasma sodium and urea concentrations.

Comment: hyponatraemia during the immediate postoperative period is a common finding, although the plasma sodium concentration rarely falls as much as it did in this case. The symptoms of confusion were probably due to the rate of fall of the plasma sodium concentration and therefore the osmolality, rather than to the actual plasma sodium concentration; the plasma osmolarity falling from approximately 296 mmol/L to 264 mmol/L. Two possible mechanisms may have contributed to this fall:

- increased ADH release causing inappropriate solute-free water retention;
- excessive infusion of hypo-osmolal fluid.

Some patients in the immediate postoperative period are prescribed up to three litres of intravenous fluid a day, some of which may be retained because of transient ADH secretion from the posterior pituitary gland. Factors which contribute to increased ADH secretion include nausea and pain and positive pressure ventilation during the operation. It is also not uncommon for five per cent dextrose to be infused in the mistaken belief that, as the patient is unable to eat, calories are being provided. One litre of five per cent dextrose provides approximately 200 kcal. The plasma urea concentration also falls because of dilution and because of the increased glomerular filtration rate associated with plasma expansion.

Diagnosis: postoperative dilutional hyponatraemia.

Disorders of potassium metabolism

CASE 4.1

A 94-year-old man was admitted to hospital with a fracture of the neck of the left femur. The only past medical history was of mild hypertension treated with a β-blocker, and a possible transient ischaemic attack, following which he took aspirin, 300 mg daily. He underwent an arthroplasty. On the day following admission, he developed mild congestive cardiac failure, and was prescribed Frumil (amiloride and frusemide). From day 6 he started to develop slight malaena and because of a falling blood haemoglobin level, he was given a blood transfusion. On day 12 he became confused; the following biochemical results were obtained:

Plasma		Reference range
Urea	17.9 mmol/L	3.0–7.0
Sodium	133 mmol/L	133–143
Potassium	6.9 mmol/L	3.6–4.6
T_{CO_2} (bicarbonate)	24 mmol/L	24–32

QUESTIONS

1. Comment on these results.
2. What other biochemical information would be useful in order to interpret the data?

1. *Comment:* there is hyperkalaemia with a slightly raised plasma urea concentration but the plasma bicarbonate concentration is normal. The hyperkalaemia is unlikely to be caused solely by renal glomerular dysfunction as it is inappropriately raised for the suggested degree of renal impairment.

2. *Additional biochemical information required to interpret the data:*

- *in vitro haemolysis,* or a delay in the separation of plasma from cells, may result in artefactual hyperkalaemia. Although severe haemolysis causes a pink discolouration of the plasma, slight haemolysis, undetectable by visual inspection, can contribute to a rise in the plasma potassium concentration;
- *previous* results would help assessment of renal function. GFR decreases with age, and most patients of this age would be expected to have a slightly high plasma urea concentration;
- *plasma creatinine concentration* should be measured to give a more accurate assessment of renal glomerular function as there are other causes of a raised plasma urea concentration (Table 2.1; p.**15**).

Repeat analysis using a fresh sample should be carried out to confirm the hyperkalaemia although, if there is electrocardiographic (ECG) evidence of a dysrhythmia, treatment should be started immediately.

Results of investigations: the plasma was not visibly haemolysed and had been separated from cells within one hour of venesection. The following results were available:

Plasma	On admission	Day 12	
Creatinine	138	150	μmol/L
Urea	11.7	17.9	mmol/L
Sodium	143	136	mmol/L
Potassium	4.7	6.9	mmol/L
T_{CO_2} (bicarbonate)	25	24	mmol/L

These results show that there was evidence of mild renal dysfunction on admission with a subsequent small rise in plasma creatinine concentration. The relatively larger rise in plasma urea concentration could be due to:

- increased catabolism of blood within the gastrointestinal tract;
- decreased GFR due to intravascular volume depletion.

QUESTION:

1. What are the possible causes of the hyperkalaemia?

1. *Some causes of hyperkalaemia* are shown in Table 4.1; these have been classified into renal and nonrenal causes. The following should be considered in this patient:

Renal causes of potassium retention

- *Renal glomerular dysfunction* causes hyperkalaemia due to a reduction in the amount of filtered sodium reaching the distal tubules, where it is exchanged for potassium. This probably contributed little in this case, since the degree of renal impairment was only slight. The fall in GFR between admission and day 12 was probably caused by a reduction in intravascular volume by a diuretic and the gastrointestinal haemorrhage.
- *Potassium-sparing diuretics* such as amiloride (a component of Frumil) cause hyperkalaemia by inhibiting distal renal tubular potassium secretion. Their use is most likely to cause hyperkalaemia in a patient with renal glomerular dysfunction. The combination of a nonsteroidal anti-inflammatory drug (NSAID), such as aspirin, may further increase the risk of hyperkalaemia as these drugs may inhibit the release of renin and therefore aldosterone.
- *Primary adrenocortical insufficiency* (Addison's disease) causes hyperkalaemia due to aldosterone deficiency. It should be considered in every patient with hyperkalaemia and uraemia, especially when accompanied by a low plasma sodium concentration. This patient did not have postural hypotension or other clinical features of advanced Addison's disease. Adrenocortical function could be investigated by performing a short tetracosactrin ('Synacthen') test (p.**180**).
- *Hyporeninaemic hypoaldosteronism,* also known as Type IV renal tubular acidosis, occurs mainly in elderly patients with mild to moderate renal impairment and is frequently associated with glucose intolerance. It is a cause of hyperchloraemic acidosis.

Table 4.1   Some causes of hyperkalaemia classified into predominant nonrenal and renal causes

Nonrenal causes	Renal causes
In vitro effect	Insufficient Na^+ available for exchange
haemolysis	renal glomerular dysfunction
delayed separation of plasma from cells	sodium depletion
leucocytosis	
thrombocytosis	Decreased Na^+/K^+ exchange
	mineralocorticoid deficiency
Increased input into ECF	hypoaldosteronism (Addison's disease)
exogenous source	congenital adrenal hyperplasia
endogenous source	hyporeninaemic hypoaldosteronism
extensive tissue cell damage	(Type IV RTA)
haemolysis	
familial hyperkalaemic	Drugs
periodic paralysis	angiotensin converting enzyme (ACE)
	inhibitors
Redistribution between cells and ECF	diuretics
acidosis	spironolactone
hypoxia	amiloride

Nonrenal cause of hyperkalaemia

- *Increased potassium input* may have occurred following the episodes of gastrointestinal haemorrhage. This is unlikely to be a major factor, since the accompanying catabolism of haemoglobin would almost certainly have resulted in a much greater rise in plasma urea concentration than seen here. Blood transfusion may also have resulted in increased potassium input, due to haemolysis, either during storage of the blood, or after transfusion.

Progress: the diuretic (Frumil) and the aspirin were stopped and he was given oral Calcium Resonium, a resin which exchanges calcium for potassium, so removing the latter into the intestinal lumen. The plasma potassium fell to within the reference range within three days.

Diagnosis: multifactorial hyperkalaemia due to the use of a potassium-'sparing' diuretic and NSAID in a patient with mild renal glomerular dysfunction and possible hyporeninaemic hypoaldosteronism.

CASE 4.2 ANOREXIA NERVOSA WITH HYPOKALAEMIA

A 32-year-old woman with a 15-year history of anorexia nervosa was referred to a gastroenterologist following a recent history of further weight loss. On examination she weighed 38.6 kg. Her blood pressure was 80/50 mmHg and the pulse was regular and 50 per minute. She had marked myotonia (impaired relaxation after muscle contraction). The following biochemical tests were performed.

Plasma			Reference range
Urea	3.8	mmol/L	3.0–7.0
Sodium	131	mmol/L	133–143
Potassium	1.9	mmol/L	3.6–4.6
Chloride	68	mmol/L	95–105
T_{CO_2} (bicarbonate)	43	mmol/L	24–32
Calcium	2.31	mmol/L	2.25–2.60
Albumin	34	g/L	35–55
Phosphate	0.54	mmol/L	0.85–1.40
Magnesium	0.60	mmol/L	0.70–1.00

Interpretation: there was a profound hypokalaemic alkalosis with mild hyponatraemia. The plasma urea concentration was within the reference range. The plasma magnesium and phosphate concentrations were low.

Comment: causes of hypokalaemia are shown in Table 4.2; possible causes in this patient include:

- *reduced dietary potassium intake;* this may be so low as to be less than the daily potassium loss, resulting in long-term negative potassium balance;
- *increased gastrointestinal potassium loss* possibly caused by vomiting (bulaemia) or taking laxatives;

Table 4.2 Some causes of hypokalaemia associated with either a metabolic alkalosis (increased plasma T_{CO_2} concentration) or a metabolic acidosis (reduced plasma T_{CO_2} concentration) (p.**42**)

Metabolic alkalosis	Metabolic acidosis
Inadequate K$^+$ intake	Renal tubular acidosis
	congenital
Drugs	acquired
diuretics:	
thiazides	Drugs
frusemide	acetazolamide
bumetanide	
steroids	Transplantation of the ureters into the
	colon or ileum or ileal loops
Mineralocorticoid excess	
hyperaldosteronism	Profound diarrhoea
primary (Conn's syndrome)	
secondary	
Cushing's syndrome	
ingestion of liquorice or carbenexolone	
congenital adrenal hyperplasia	
Bartter's syndrome (p.42)	
Miscellaneous	
hypercalcaemia	
hypomagnesaemia	

- *increased urinary potassium loss* if the patient also abuses diuretics and if there is significant volume depletion and hypochloraemia following vomiting. Volume depletion stimulates secondary hyperaldosteronism and hypochloraemia is associated with a reduction in proximal renal tubular reabsorption of sodium and with an increased delivery of sodium to the distal renal tubules where it can be exchanged for either potassium or hydrogen ions.

Inadequate mineral intake may have contributed to hypophosphataemia and hypomagnesaemia and these may have been exacerbated by vomiting as this patient almost certainly had bulaemia.

Diagnosis: anorexia nervosa with severe hypokalaemia.

CASE 4.3 ADRENOCORTICAL INSUFFICIENCY (ADDISON'S DISEASE)

A 45-year-old woman was found collapsed at home. She had attended the outpatient clinic about three weeks previously and had been noted as being well. She had a history of Cushing's disease, diagnosed 10 years previously for which she had had a bilateral adrenalectomy followed by steroid replacement with hydrocortisone and fludrocortisone.

Plasma		Reference range
Urea	17.3 mmol/L	2.5–7.0
Sodium	116 mmol/L	135–145
Potassium	5.9 mmol/L	3.5–4.8
T_{CO_2} (bicarbonate)	15 mmol/L	22–32

Interpretation: hyperkalaemic acidosis with hyponatraemia and a moderately raised plasma urea concentration.

Comment: hyperkalaemia associated with hyponatraemia is suggestive of mineralocorticoid deficiency. The metabolic acidosis (reduced plasma T_{CO_2} concentration) could possibly have been caused by renal glomerular impairment secondary to intravascular volume depletion (raised plasma urea concentration); however, the plasma potassium concentration is inappropriately raised for the degree of acidosis and renal impairment. The low plasma sodium concentration in the presence of a moderately raised plasma urea concentration is suggestive of intravascular volume depletion with excess water retention.

Progress: further enquiry revealed that she had 'run out' of tablets three days previously. Hydrocortisone and fludrocortisone were restarted and within two days the plasma urea and electrolyte concentrations had returned to normal.

Plasma	On admission	Day 1	Day 2	
Urea	17.3	8.1	4.6	mmol/L
Sodium	116	127	137	mmol/L
Potassium	5.9	3.4	3.8	mmol/L
T_{CO_2} (bicarbonate)	15	18	23	mmol/L

Diagnosis: profound mineralocorticoid deficiency (Addisonian crisis) because the patient had discontinued steroid replacement.

CASE 4.4   ECTOPIC ADRENOCORTICOTROPHIC HORMONE (ACTH) SECRETION WITH HYPOKALAEMIA

A 69-year-old man was admitted to hospital for investigation of haemoptysis. There was a history of recent weight loss; he smoked 30 cigarettes per day.

Plasma		Reference range
Urea	7.9 mmol/L	3.0–7.0
Sodium	143 mmol/L	133–143
Potassium	2.5 mmol/L	3.6–4.6
T_{CO_2} (bicarbonate)	38 mmol/L	24–32

Interpretation: hypokalaemic alkalosis with a plasma sodium concentration at the upper limit of the reference range.

Comment: the biochemical findings are compatible with a markedly excessive mineralocorticoid activity. Mineralocorticoids enhance the reabsorption of sodium from the distal renal tubules in exchange for either potassium or hydrogen ions (H^+). As the intracellular concentration of potassium falls H^+ ions are preferentially exchanged. As H^+ ions enter the luminal fluid in exchange for Na^+ ions, the bicarbonate ions generated during H^+ secretion enter the circulation along with Na^+ ions (Fig. 4.1), thus contributing to the metabolic alkalosis normally associated with a reduction in total body potassium. As the plasma sodium concentration rises, so does the osmolality, which stimulates ADH release, thus potentiating the absorption of water from the collecting tubules and limiting the development of hypernatraemia.

The clinical details and biochemical findings are suggestive of a carcinoma of the lung with ectopic ACTH production. In this condition there are frequently profound metabolic disturbances with minimal clinical evidence of Cushing's syndrome.

Diagnosis: ectopic ACTH secretion from a small-cell anaplastic (oat cell) carcinoma of the lung, causing a profound hypokalaemic alkalosis.

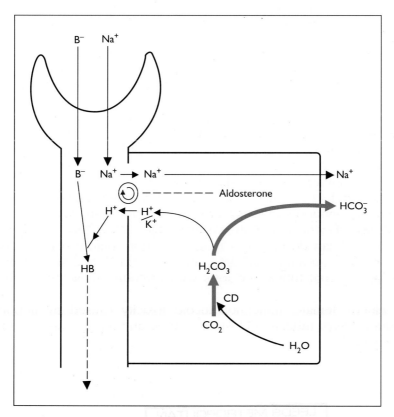

Fig. 4.1 The generation of bicarbonate and the production of a metabolic alkalosis in total body potassium depletion.
(B^- = nonbicarbonate base, CD = carbonate dehydratase.)

CASE 4.5 TWO CASES OF PROFOUND HYPOKALAEMIA

Two patients were referred to hospital within six weeks of each other. The following investigations were performed.

Patient	I	2	
Gender	female	male	
Age	70	39	
History	Generalized weakness recent onset	Transferred from another hospital for management of carcinoma of the pancreas; no clinical notes available	
Plasma			**Reference range**
Urea	3.9	5.0 mmol/L	2.5–8.0
Sodium	147	137 mmol/L	133–143
Potassium	1.4	1.5 mmol/L	3.6–4.6
T_{CO_2} (bicarbonate)	42	21 mmol/L	24–32

Comment: although both patients had severe hypokalaemia, the differential diagnosis in each, based on the initial biochemical tests, was different.

The elderly female had a hypokalaemic alkalosis with an unequivocally raised plasma sodium concentration. This pattern is compatible with excessive mineralocorticoid activity, the causes of which are shown in Table 4.2 (p.**39**). She was normotensive and there were no cushingoid features. The possibility of Bartter's syndrome was considered initially. This extremely rare condition is caused by hyperplasia of the juxtaglomerular apparatus with biochemical findings of hyperaldosteronism except that patients are normotensive. It is unlikely, however, to present for the first time at this age. Within hours of being admitted she was seen buying liquorice sweets from the hospital shop. Liquorice contains glycyrrhizinic acid which has a mineralocorticoid-like effect.

The male patient had a similar plasma potassium concentration; however it was associated with a low-normal plasma T_{CO_2} concentration, suggestive of a metabolic acidosis, causes of which are shown in Table 4.2 (p.**39**). He had received chemotherapy and had developed a fungal infection. Before being transferred he had been started on an antifungal drug, amphotericin B, and subsequently developed an acquired renal tubular acidosis with profound renal potassium loss.

Diagnosis: elderly female: mineralocorticoid toxicity caused by liquorice. Male patient: acquired renal tubular acidosis with profound renal potassium loss caused by amphotericin B.

Disorders associated with altered blood pressure

CASE 5.1

A 73-year-old woman with a 12-year history of moderately well-controlled hypertension was admitted for a hysterectomy. Shortly after induction of anaesthesia, systolic blood pressure rose rapidly from 120 mmHg to 230 mmHg. ECG changes were consistent with extensive myocardial ischaemia.

Over the next three days her cardiovascular condition stabilized, but on several occasions she was noted to be peripherally cold and clammy. On further questioning she revealed that she had been suffering from headaches and episodes of profuse sweating in recent months. The results of the following investigations were obtained one week after the operation:

Plasma		Reference range
Creatinine	124 μmol/L	55–110
Urea	10.6 mmol/L	2.5–7.0
Sodium	137 mmol/L	135–145
Potassium	3.6 mmol/L	3.5–4.8
Calcium	2.60 mmol/L	2.15–2.55
Albumin	34 g/L	27–42
Phosphate	0.77 mmol/L	0.60–1.40
Glucose	9.0 mmol/L	3.6–6.5

QUESTIONS

1. What are the causes of hypertension?
2. How do you interpret the results of the above tests and how do they contribute to the diagnosis?
3. What additional biochemical tests should be requested to diagnose the cause of hypertension in this patient?

1. *Causes of hypertension:* essential hypertension, raised blood pressure of unknown cause, accounts for approximately 95 per cent of cases. Some causes of secondary hypertension are shown in Table 5.1.

A possible cause in this patient is a phaeochromocytoma, the characteristic features of which are shown in Table 5.2. This patient presented with paroxysmal hypertension, headaches and sweating. Phaeochromocytoma accounts for less than one per cent of cases.

2. *Interpretation of initial biochemical tests:* the plasma urea and creatinine concentrations were marginally raised indicating mild renal glomerular dysfunction. This could be a consequence of long-standing hypertension, but was not severe enough to be its cause. The plasma potassium concentration was normal and the plasma sodium concentration was at the lower end of the reference range; this combination of test results excluded excess mineralocorticoid activity, making the diagnosis of either Cushing's or Conn's syndrome unlikely.

There was mild hypercalcaemia, with a low phosphate concentration relative to the GFR. This is characteristic of primary hyperparathyroidism; the diagnosis can be confirmed by relating the plasma calcium concentration to that of PTH although this is rarely necessary. Primary hyperparathyroidism is associated with an increased incidence of hypertension.

The random plasma glucose concentration was slightly raised, indicating either impaired glucose tolerance or mild diabetes mellitus. Diabetic nephropathy is a cause of hypertension. Impaired glucose tolerance or frank diabetes mellitus may also occur in Cushing's syndrome or phaeochromocytoma.

Table 5.1   Some causes of secondary hypertension

Renal disease
    chronic renal glomerular dysfunction
    polycystic kidney
    renovascular disease
        renal artery stenosis
Endocrine disease
    diabetes mellitus
    Cushing's syndrome
    primary hyperaldosteronism (Conn's syndrome)
    phaeochromocytoma
    primary hyperparathyroidism
Drug-induced
    oral contraceptives
    nonsteroidal anti-inflammatory drugs (NSAID)
    steroids

Table 5.2   Characteristic clinical features of a phaeochromocytoma

Severe or malignant hypertension
Hypertension resistant to treatment
Paroxysmal hypertension
Hypertension in children or young adults
Characteristic symptoms
    headache
    perspiration
    palpitations

3. *Additional biochemical tests to diagnose the cause of hypertension:* a 24-hour urine sample should be collected for the measurement of free catecholamines (adrenaline, noradrenaline and dopamine) or their metabolites (metadrenalines or 4-hydroxy 3-methoxy mandelate [HMMA]). This is preferable to the measurement of plasma adrenaline or noradrenaline, which may give false positive results due to the physical and mental stress of venesection, or false negative results in an episodically secreting tumour. Twenty-four-hour excretion up to twice the upper reference limit may occur in samples from patients with essential hypertension.

The following results were obtained from a 24-hour urine collection:

Urine		Reference range
HMMA	215 μmol/24 hour	9–31
Noradrenaline	3387 nmol/24 hour	<500
Adrenaline	3963 nmol/24 hour	<100
Dopamine	5763 nmol/24 hour	<2700

QUESTIONS

1. Comment on these results.
2. What additional investigations are indicated to explain some of the other biochemical abnormalities and to locate the site of the tumour?

1. *Comment:* both urinary HMMA and free catecholamine concentrations are significantly elevated, almost certainly due to a phaeochromocytoma.

2. *Additional investigations:* the coexistence of primary hyperparathyroidism with a phaeochromocytoma raises the possibility of multiple endocrine neoplasia (MEN) Type II. However, this is unlikely since the majority of cases of MEN II include medullary carcinoma of the thyroid and present in the third or fourth decade of life. Plasma parathyroid hormone (PTH) and calcitonin concentrations were measured to confirm a diagnosis of primary hyperparathyroidism and to exclude medullary carcinoma of the thyroid.

Plasma		Reference range
PTH	36 pmol/L	10–50
Calcitonin	0.06 µg/L	<0.08

Although the plasma PTH concentration is within the reference range, it is inappropriately raised in the presence of hypercalcaemia and so confirms primary hyperparathyroidism; the diagnosis of medullary carcinoma of thyroid and therefore MEN II are highly unlikely.

Imaging studies were performed to localize the tumour:

- *abdominal CT scan* demonstrated a 3 cm mass in the left adrenal. The right adrenal appeared normal and there was no evidence of lymph node enlargement or hepatic involvement (Fig. 5.1)
- ^{131}*I-metaiodobenzylguanidine (MIBG) scan* demonstrated immediate intense uptake of radioactivity at the upper pole of the left kidney. There was no significant uptake elsewhere, confirming the presence of a unilateral catecholamine secreting tumour (Fig. 5.2).

Progress: a large tumour was successfully removed from the left adrenal gland. This contained clusters of densely packed cells giving a positive chromaffin

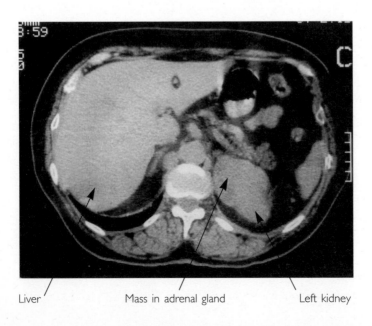

Liver

Mass in adrenal gland

Left kidney

Fig. 5.1 Computerized tomography of the abdomen demonstrating a mass in the left adrenal gland, which was subsequently shown to be a phaeochromocytoma

Immediate uptake

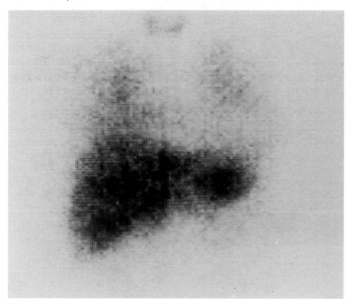

Fig. 5.2 131I-metaiodobenzyl-guanidine (MIBG) scan demonstrating immediate intense uptake by a large mass at the upper pole of the left kidney and also by the liver. The picture taken at 24 hours identifies the mass more clearly (note the normal uptake by the thyroid gland). These findings supported the diagnosis of a phaeochromocytoma

At 24 hours

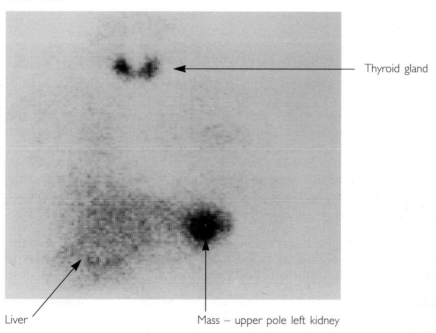

Thyroid gland

Liver

Mass – upper pole left kidney

reaction (Fig. 5.3). There was no evidence of malignancy. One week after surgery, urinary HMMA excretion was 19 μmol/24 hour, indicating successful removal of the catecholamine secreting tissue.

Diagnosis: phaeochromocytoma and primary hyperparathyroidism.

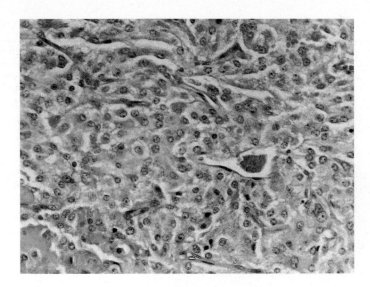

Fig. 5.3 Haemotoxylin and eosin (H and E) stained section (× 400 magnification) of a phaeochromocytoma taken from the adrenal gland. The tumour cells show vesicular nuclei with a stippled chromatin pattern, inconspicuous nucleoli and abundant eosinophilic cytoplasm; numerous dilated vessels are seen within the tumour. The tumour stained positively for the neuroendocrine marker, chromogranin

CASE 5.2   CHRONIC INFLAMMATORY BOWEL DISEASE (CROHN'S DISEASE)

A 62-year-old man, with a 25-year history of chronic inflammatory bowel disease (Crohn's disease), was admitted to hospital for assessment. He was passing five loose bowel motions per day but there was no mucus or blood in the stool. He had lost approximately 5 kg in weight during the past six months. Twenty-five years previously he had a total colectomy and ileorectal anastomosis. He subsequently developed an ileal stricture for which he had a partial resection of the ileum.

On examination he had marked finger clubbing. His blood pressure was 90/40 mmHg lying with a significant postural drop to 75/30 mmHg. Resting pulse was 86 per minute. The following biochemical investigations were performed.

Plasma		**Reference range**
Creatinine	217 μmol/L	75–120
Urea	43.6 mmol/L	3.0–8.0
Sodium	120 mmol/L	133–143
Potassium	4.8 mmol/L	3.6–4.6
T_{CO_2} (bicarbonate)	13 mmol/L	22–32
Blood		
Haemoglobin (Hb)	12.6 g/dl	13.0–18.0
White blood cell count (WBC)	8.7 ×10⁹/L	4.0–11.0
Erythrocyte sedimentation rate (ESR)	63 mm in one hour	3–5

Comment: the plasma urea concentration was approximately five times the upper adult reference limit, whereas that of creatinine was only twice the upper reference limit. This was suggestive of a prerenal uraemia due to either intravascular volume depletion and reduced GFR, or an abnormally high supply of amino nitrogen, such as that derived from excessive protein catabolism, which is then used for urea synthesis (Table 2.1; p.**15**). The plasma sodium concentration was low and that of potassium at the upper end of the reference range.

Differential diagnosis includes:

*   marked intravascular volume depletion with a prerenal uraemia;
*   adrenocortical insufficiency (Addison's disease).

Further investigations: there is an association between chronic inflammatory bowel disease and amyloidosis of the adrenal gland although it rarely causes marked adrenal insufficiency. However, because of the clinical findings of marked postural hypotension together with the biochemical findings, adrenal insufficiency must be excluded. The results of a short tetracosactrin ('Synacthen') test (p.**180**) are shown below.

Short tetracosactrin test

Time	0	30	45 minutes
Cortisol	378	694	812 nmol/L

Interpretation: normal response. Following 250 μg 'Synacthen', the plasma cortisol concentration should increase by at least 200 nmol/L to a concentration greater than 550 nmol/L by 30 minutes.

In the presence of hypovolaemia, hyponatraemia is usually a consequence of either renal or gastrointestinal sodium loss. Therefore, urinary and faecal sodium concentrations were measured.

Urine

Sodium	<1	mmol/L

Faeces

Volume	1.26	L/24 hours
Sodium	81	mmol/L

Interpretation: there was excessive faecal water and sodium loss. The low spot urinary sodium concentration indicated a normal renal tubular response to the secondary hyperaldosteronism caused by hypovolaemia. Profound hypovolaemia stimulates ADH production with excessive water reabsorption from the renal collecting ducts causing a dilutional hyponatraemia. Faecal sodium loss is normally reduced as a consequence of secondary hyperaldosteronism because of enhanced sodium absorption from the colon; however, this patient had had a total colectomy and, because of impaired sodium absorption, excessive amounts were lost in the stool.

Management: the patient was given four litres per day of isotonic saline with oral fluids initially. Over the next five days his weight increased and the blood pressure rose to 110/70 mmHg lying, 80/50 mmHg standing. There were no symptoms or signs of fluid overload. Repeat biochemical investigations are shown below.

Plasma	On admission	Day 6	
Creatinine	217	203	μmol/L
Urea	43.6	14.1	mmol/L
Sodium	120	136	mmol/L
Potassium	4.8	4.3	mmol/L

Diagnosis: faecal sodium loss with postural hypotension due to chronic inflammatory bowel disease (Crohn's disease).

CASE 5.3   PRIMARY HYPERALDOSTERONISM (CONN'S SYNDROME)

A 26-year-old woman developed mild hypertension during her first pregnancy. Six weeks following delivery, her blood pressure remained elevated at 160/100 mmHg. Preliminary investigation revealed the following:

Plasma		Reference range
Urea	4.8 mmol/L	2.5–7.0
Sodium	142 mmol/L	135–145
Potassium	3.1 mmol/L	3.5–4.8
T_{CO_2} (bicarbonate)	36 mmol/L	22–32

Interpretation: hypokalaemic alkalosis with a high-normal plasma sodium concentration.

Comment: the biochemical findings, in association with mild to moderate hypertension, are suggestive of excessive mineralocorticoid secretion or intake. There were no clinical features suggestive of Cushing's disease. Although uncommon the possible diagnosis of primary hyperaldosteronism was considered.

Primary hyperaldosteronism may be caused by:

- a single adenoma within the zona glomerulosa of one of the adrenal glands in about half the cases;
- multiple adenomata in about 10 per cent of cases;
- bilateral nodular hyperplasia in most of the remaining cases.

There is partial autonomy of aldosterone secretion with suppression of renin synthesis by the juxtaglomerular apparati within the kidneys. Aldosterone enhances sodium, and therefore water, reabsorption from the distal renal tubules in exchange for either hydrogen or potassium ions, resulting in intravascular volume expansion and hypokalaemic alkalosis.

To make the diagnosis of primary hyperaldosteronism, the patient must not be sodium depleted and must be on a diet containing approximately 100 mmol of sodium per day. Hypokalaemia should be corrected by giving potassium supplements. After eight hours recumbency a blood sample is taken and followed by two further samples after half an hour and four hours when the patient has been up and walking around (p.**182**) *(the local protocol for investigating such patients should be consulted).*

The following results were obtained in this patient:

Plasma	Supine	Erect 30 minutes	4 hours	
Aldosterone	690 (100–450)		630	pmol/L
Renin activity	< 0.2 (1.1–2.7)	< 0.2 (2.8–4.5)		pmol/ml/hour

Comment: although the plasma aldosterone concentrations were only moderately elevated they were inappropriately raised in the presence of suppressed plasma renin activity, thus confirming primary rather than secondary hyperaldosteronism.

Further investigations: selective venous catheterization was performed to identify the site of excessive aldosterone production. In addition to the measurement of plasma aldosterone, plasma cortisol concentrations were assayed on specimens taken through the catheter, to help confirm the position of its tip at the time of sampling.

Site of catheter	Aldosterone (A) (pmol/L)	Cortisol (C) (nmol/L)	A/C ratio
Left adrenal vein	94500	5400	17.5
Right adrenal vein	5200	13500	0.4
Lower inferior vena cava	2140	790	2.7
Peripheral vein	1785	750	2.4

Interpretation: there was a high plasma aldosterone concentration and a high A/C ratio in the sample obtained from the left adrenal vein, suggesting that the source of excess aldosterone synthesis was from the left adrenal gland.

Diagnosis: primary hyperaldosteronism (Conn's syndrome).

CASE 5.4   COMPLICATIONS OF TREATMENT OF HYPERTENSION

A 45-year-old man, with a long history of hypertension, attended his general practitioner (GP) for an annual health check. He was obese; his blood pressure was moderately well controlled on a thiazide diuretic, bendrofluazide 5 mg daily. Because glycosuria was detected by 'dip-stick' testing of the urine, his GP arranged for the following investigations to be performed after a 12-hour fast.

Plasma			Reference range
Urea	9.5	mmol/L	3.0–8.0
Sodium	135	mmol/L	134–144
Potassium	2.9	mmol/L	3.5–4.8
T_{CO_2} (bicarbonate)	34	mmol/L	24–32
Glucose	15.4	mmol/L	2.2–5.5

Serum			Desirable range
Total cholesterol	7.2	mmol/L	<6.5
Triglyceride	3.2	mmol/L	0.5–2.0
HDL-cholesterol	0.7	mmol/L	>0.9

Interpretation: there was a hypokalaemic alkalosis (raised plasma Tco$_2$ concentration). The plasma urea concentration was slightly raised, suggestive of either intravascular volume depletion with a prerenal uraemia or renal glomerular dysfunction. The fasting plasma glucose concentration was significantly raised; however diabetes mellitus should not be diagnosed until a second test has confirmed a raised plasma glucose concentration (p.**173**). There was a mixed hyperlipidaemia with a low serum HDL-cholesterol concentration.

Comment: this patient had developed some of the metabolic complications of treatment with thiazide diuretics. These include:

- *hypokalaemic alkalosis.* Thiazide diuretics inhibit sodium, and therefore water reabsorption from the cortical segments of the loops of Henle. Increased delivery of sodium to the distal renal tubules, associated with aldosterone enhanced exchange for either potassium or hydrogen ions, may result in hypokalaemia and a metabolic alkalosis;
- *impaired glucose tolerance* or frank diabetes mellitus, particularly if there are other predisposing conditions present, such as obesity;
- *mixed hyperlipidaemia,* although obesity and impaired glucose tolerance may also have been contributing factors for secondary hyperlipidaemia;
- *hyperuricaemia* as thiazide diuretics inhibit renal urate excretion.

Increasing the dose of thiazide diuretics in the management of hypertension has little effect on blood pressure control but may significantly increase the incidence of metabolic complications.

Diagnosis: thiazide diuretic-induced hypokalaemic alkalosis, glucose intolerance and combined hyperlipidaemia.

CASE 5.5 ADDISON'S DISEASE DUE TO INADEQUATE MINERALOCORTICOID REPLACEMENT

A 41-year-old woman was diagnosed as having Addison's disease following a prolonged period of lassitude and weight loss, associated with increasing skin pigmentation and postural hypotension. She was started on steroid hormone replacement therapy (hydrocortisone and fludrocortisone twice daily). Three months later, although her symptoms had greatly improved, she still complained of feeling dizzy when she stood up; her blood pressure was 100/75 mmHg lying and 90/65 mmHg standing. She was admitted to the day ward in order to assess the adequacy of the steroid hormone replacement.

Plasma				Reference range
Creatinine		82	μmol/L	55–110
Urea		7.2	mmol/L	3.0–7.0
Sodium		134	mmol/L	133–143
Potassium		4.7	mmol/L	3.6–4.6
TcO$_2$ (bicarbonate)		29	mmol/L	24–32
Renin				
(pre-dose)	09.00	4.36	pmol/ml/hour	1.1–2.7
Cortisol				
(pre-dose)	09.00	78	nmol/L	
	11.00	1120	nmol/L	
	13.00	703	nmol/L	
	15.00	460	nmol/L	
(pre-dose)	16.00	149	nmol/L	
	18.00	630	nmol/L	
	20.00	404	nmol/L	

Interpretation: the slightly raised plasma urea concentration with a normal plasma creatinine concentration was suggestive of intravascular volume depletion with a prerenal uraemia; this finding was consistent with the raised plasma renin activity, and supported the clinical finding of postural hypotension. The marginally raised plasma potassium concentration, with the sodium concentration at the lower end of the reference range, also supported the suggestion of inadequate mineralocorticoid replacement. Although the peak plasma cortisol concentration, at 11.00 hours, was slightly high the mean concentration was within the expected reference range and, in particular, the trough concentrations (pre-dose) were low but not too low.

Comment: the cause for this patient's Addison's disease was autoimmune destruction of the adrenal cortex resulting in loss of secretion of both cortisol (a glucocorticoid) and aldosterone (a mineralocorticoid). Most patients with Addison's disease require both glucocorticoid and mineralocorticoid replacement. By giving hydrocortisone in divided doses with approximately two-thirds in the morning, it is possible to achieve a normal diurnal variation, similar to that in the nonstressed, well individual.

If the mineralocorticoid replacement, and therefore plasma activity, is too low, reduced distal renal tubular sodium reabsorption results in increased urinary sodium and water loss, intravascular volume depletion and therefore postural hypotension. Reduced potassium secretion at the same site results in hyperkalaemia. Measurement of the plasma renin activity has been used to assess the degree of intravascular volume depletion as reduced renal blood flow increases renin release from the juxtaglomerular apparatus in the kidneys. The high pre-dose resting plasma renin activity confirmed the clinical finding of postural hypotension was due to intravascular volume depletion. However, this investigation is rarely necessary. The dose of fludrocortisone was increased.

Diagnosis: inadequate mineralocorticoid replacement in a patient being treated for Addison's disease.

Endocrine disorders

CASE 6.1

A 62-year-old man consulted his general practitioner; during the previous six months he had experienced severe headaches and excessive perspiration and had become aware that he had developed tunnel vision. He had a 15-year history of hypertension, but had otherwise only suffered from vague aches and pains in his shoulder and hand joints, which had been diagnosed as osteoarthrosis.

On examination he had coarse facial features with a prominent jaw. There was bitemporal hemianopia. His blood pressure was 145/105 mmHg. Glycosuria was detected.

Because of the visual field defect, a lateral skull X-ray was performed. This demonstrated a grossly enlarged pituitary fossa, consistent with an intrasellar mass.

The following initial biochemical investigations were carried out:

Plasma			Reference range
Glucose	8.6	mmol/L	3.5–6.5
Luteinizing hormone (LH)	2.7	U/L	0.7–6.0
Follicle stimulating hormone (FSH)	5.1	U/L	<6
Prolactin	281	mU/L	<425
Testosterone	15.1	nmol/L	10.0–35.0
Growth hormone	8.0	mU/L	<10
Cortisol (random)	427	nmol/L	250–700
Thyrotrophin-stimulating hormone (TSH)	2.3	mU/L	0.3–6.0
Free T_4	13.2	pmol/L	9.4–25.0

QUESTIONS

1. What is the most likely diagnosis on the basis of the clinical and radiological findings?
2. Why were the above investigations performed and how would you interpret the results?
3. What further biochemical tests are indicated?

1. *The most likely diagnosis.* The history of headaches, the finding of bitemporal hemianopia, and the radiological evidence of enlargement of the pituitary fossa on the skull X-ray strongly suggest a pituitary tumour extending upwards out of the sella turcica and compressing the optic chiasma.

A growth-hormone (GH) secreting acidophil adenoma is most likely in view of the characteristic facial appearance, osteoarthrosis, sweating and hypertension. He also has glycosuria and a slightly raised plasma glucose concentration. These are all features of acromegaly (Table 6.1).

Table 6.1 Some clinical features of acromegaly

Effects of growth hormone excess
    Enlargement of hands and feet
    Enlargement of the tongue and jaw (prognathism)
    Coarse facies and thickening of the skin
    Excessive perspiration
    Degenerative arthritis (osteoarthrosis)
    Hypertension
    Impaired glucose tolerance
Effects of tumour growth
    Headaches
    Visual field defects
    Hyperprolactinaemia
    Cranial nerve palsies
    Panhypopituitarism

2. *The biochemical investigations were requested* to confirm excessive growth hormone secretion and to identify evidence of hypopituitarism (reduced secretion of other anterior pituitary hormones) due to destruction of pituitary tissue by a tumour.

The plasma GH concentration was within the reference range. However, GH secretion is pulsatile and concentrations normally fall to very low or undetectable levels between pulses. In acromegaly, GH secretion loses its pulsatile nature; although the serum GH concentration at any particular time of day may still be within the reference range, the overall secretion rate is increased. GH secretion is normally inhibited by a rise in the plasma glucose concentration. GH secretion by a pituitary adenoma is autonomous and not subject to this control. Failure of serum GH concentrations to fall during a glucose load is diagnostic of acromegaly.

Plasma luteinizing hormone (LH), follicle-stimulating hormone (FSH), testosterone, TSH and free T_4, cortisol and prolactin concentrations were all normal, indicating that hypopituitarism is unlikely. In hypopituitarism, the concentrations of both pituitary hormones and the hormones produced by their target organs are low. However, the serum prolactin concentration may rise if there is disruption of the pituitary stalk, since prolactin is normally under inhibitory control from the hypothalamus. If the pituitary hormone concentrations had been low or low-normal in the presence of normal or low testosterone, free-T_4 and cortisol concentrations, then pituitary responsiveness to stimulation should have been tested using a combined pituitary stimulation test with insulin, thyrotrophin-releasing hormone (TRH) and gonadotrophin-releasing hormone (GnRH) (p.**176**).

3. *Further biochemical investigations.* A glucose load (tolerance) test was performed (p.**182**).

Time (minutes)	Glucose (mmol/L)	GH (mU/L)
0 (fasting)	7.7	9
30	12.4	12
60	15.7	12
90	13.3	10
120	10.2	10

Interpretation: serum GH concentrations failed to fall in response to a rise in the plasma glucose concentration. They normally fall to less than 2 mU/L during a glucose load. In this case, there was in fact a small rise in GH concentrations. This paradoxical rise is often found in acromegaly. The high plasma glucose concentration at 120 minutes indicates impaired glucose tolerance. GH excess causes insulin resistance; impaired glucose tolerance may develop in approximately 25 per cent of cases with such excess.

The serum insulin-like growth factor (IGF-1) concentration was measured.

Serum		Reference range
IGF-1	75 nmol/L	4–33

Interpretation: many of the effects of GH are mediated through IGF-1, which is synthesized in the liver. GH stimulates IGF-1 production but, unlike GH, plasma IGF-1 concentrations are nonpulsatile. Serum IGF-1 concentrations, therefore, reflect overall GH secretion, and the raised serum concentrations indicate excessive GH production. Measurement of the serum IGF-1 concentration is a useful adjunct to diagnose acromegaly, although increased concentrations may occur at puberty and during pregnancy; serum concentrations fall due to malnutrition or following prolonged fasting.

The pituitary–adrenal axis was tested further by carrying out a short tetracosactrin test (p.**180**).

Short tetracosactrin ('Synacthen') test

Time	0	30	60 minutes
Cortisol	315	501	639 nmol/L

Interpretation: following 250 μg 'Synacthen', there was a delayed but adequate response in the plasma cortisol concentration. This delay was possibly caused by chronic understimulation of the adrenal cortex. If the response is equivocal and the cause unknown, then either a five-hour or three-day tetracosactrin test should be performed (p.**181**). It is important to establish that adrenal reserve is normal in a patient with possible panhypopituitarism who is about to undergo the stress associated with surgery.

A magnetic resonance imaging (MRI) scan demonstrated a large mass occupying and expanding the region of the pituitary fossa and extending into the sphenoid sinus. In view of the risk of further damage to adjacent nervous tissue, he was referred immediately for surgery.

Diagnosis: acromegaly with impaired glucose tolerance.

CASE 6.2 CARCINOMA OF THE ADRENAL GLAND – EXCESS ANDROGEN SECRETION

A 34-year-old woman presented to her general practitioner with a six-month history of increased hair growth on the upper lip and eyebrows, in the axillae and pubic area associated with temporal recession of the scalp hairline. She also complained of general malaise and increasing weight.

On examination, she had truncal obesity, increased hair growth as described and abdominal striae. She had mild hypertension, with a blood pressure of 130/95 mmHg. There were no visual field defects. No abdominal masses were palpable. She was clinically euthyroid.

The following initial investigations were carried out:

Serum			Reference range
LH	1.6	U/L	3.0–8.0
FSH	4.4	U/L	1.0–9.0
Testosterone	4.3	nmol/L	0.3–2.5
Cortisol (09.00 hours)	620	nmol/L	150–700
Free-T$_4$	13.2	pmol/L	9.4–25.0
TSH	0.8	mU/L	0.3–6.0

Interpretation: serum total testosterone concentration is raised. Serum luteinizing hormone (LH) concentration is low because of normal feedback suppression by testosterone. The 09.00 hours serum cortisol concentration is within the early morning reference range.

Comment: increased hair growth (hirsutism) was the presenting complaint (Table 6.2). Additional clinical features suggestive of masculinization (virilism) include:

- temporal hair recession;
- increased hair growth in male distribution;
- cliteromegaly;
- breast atrophy;
- deepening voice.

Table 6.2 Some causes of hirsutism

Familial or racial
Polycystic ovary syndrome
Congenital adrenal hyperplasia
    late onset 21α-hydroxylase deficiency
Androgen-producing adrenal or ovarian carcinoma
Associated clinical conditions
    hypothyroidism
    obesity
Androgen treatment

Table 6.3 Some causes of virilism

Ovarian tumours
   arrhenoblastoma
   hilus-cell tumour
Adrenocortical disorders
   tumours, usually carcinoma
   pituitary-dependent Cushing's disease (rare)
   congenital adrenal hyperplasia

Virilism is uncommon (Table 6.3) but much more serious than hirsutism; it is always associated with significantly increased androgen secretion.

There were some clinical features suggestive of Cushing's syndrome. However, the diagnostic value of a single serum cortisol concentration in excluding the diagnosis is limited because of the diurnal variation of cortisol secretion and because it may be inappropriately low for the degree of stress. Therefore, this result does not exclude hypercortisolism. Initially, either a 24-hour urinary free cortisol excretion should have been measured or an overnight dexamethasone suppression test performed (p.**179**).

Further investigations: these were performed to determine whether the excess andro-gen secretion was of ovarian or adrenal origin, and secondly to determine the cause.

The following biochemical investigations were carried out:

Serum			**Reference range**
Dehydroepiandrosterone sulphate (DHAS)	15.8	µmol/L	1.7–11.5
Androstenedione	9.8	nmol/L	4.0–10.2
17α-hydroxyprogesterone	15.2	nmol/L	<20
Cortisol 09.00 hours	733	nmol/L	150–700
midnight	677	nmol/L	<250
ACTH 09.00 hours	<7	ng/L	<25

Urine

Free cortisol	2880 nmol/24 hours	<300

Dynamic tests

Overnight dexamethasone suppression test (2 mg)

Cortisol 09.00 hours	705	nmol/L	<190

High-dose dexamethasone suppression test (2mg six-hourly for 48 hours)
Cortisol

Day 1 09.00 hours	854	nmol/L	
Day 3 09.00 hours	777	nmol/L	

Comment: the interpretation of these investigations is summarized in Table 6.4. The raised serum DHAS concentration indicates an adrenal source of the exces-sive androgen secretion; the ovary does not synthesize DHAS. Congenital adrenal hyperplasia is unlikely as the serum 17α-hydroxyprogesterone concentration is normal.

There is loss of diurnal rhythm of cortisol secretion, characteristic of Cushing's syndrome. However, this is an unreliable test as other conditions, such as stress and obesity, may cause a physiological increase in cortisol secretion. The high

Table 6.4 Interpretation of biochemical tests in the diagnosis of hirsutism

Plasma findings					Clinical diagnosis
Testosterone	DHAS	LH	FSH	17α-OH progesterone	
N or slightly ↑	N or slightly ↑	N	N	N	Simple hirsutism
↑	N	↑	N or ↓	N	Polycystic ovary syndrome
↑↑	N	N or ↓	N or ↓	N	Ovarian tumour
N or slightly ↑	↑↑	N or ↓	N or ↓	N	Adrenocortical tumour
↑	↑	N or ↓	N or ↓	↑	Congenital adrenal hyperplasia

N = normal; ↑ = increase; ↓ = decrease

24-hour urinary free cortisol excretion indicates excessive cortisol secretion. This investigation may also be affected by stress. Failure of cortisol suppression after overnight dexamethasone indicates impaired feedback regulation of cortisol synthesis. Both investigations are useful 'first line' tests to diagnose Cushing's syndrome; however, it is rarely necessary to perform both but, if they are, it is important that the 24-hour urine collection is completed before dexamethasone is given.

Failure of serum cortisol concentration to fall after the high dose of dexamethasone excludes the diagnosis of Cushing's disease (bilateral adrenal hyperplasia secondary to a pituitary adenoma). The serum ACTH concentration is very low in the presence of a raised serum cortisol concentration indicating normal feedback suppression, by cortisol, of pituitary ACTH release. This excludes ectopic, and therefore unregulated, ACTH production.

These results indicate excess cortisol and androgen production of adrenal origin. This could be caused by an adrenal adenoma or carcinoma; the significantly increased serum androgen concentration suggest that it is more likely to be a carcinoma. Computerized tomography of the abdomen demonstrated a mass in the left adrenal gland. The patient underwent an exploratory laporotomy and a tumour was removed.

Diagnosis: adrenal carcinoma.

CASE 6.3 POSTPARTUM PANHYPOPITUITARISM (SHEEHAN'S SYNDROME)

A 34-year-old woman had a postpartum haemorrhage after a normal pregnancy and delivery. Subsequently she failed to establish lactation and developed anorexia, nausea and dizziness on standing. The only abnormal finding on clinical examination was postural hypotension with a standing blood pressure of 90/60 mmHg. The following investigations were performed.

Serum			Reference range
LH	<0.5	U/L	3.0–8.0
FSH	1.7	U/L	1.0–9.0
Oestradiol	<100	pmol/L	110–180
Prolactin	593	U/L	<700
Free-T$_4$	7.1	pmol/L	10–20
TSH	0.8	mU/L	0.3–6.0

Interpretation: the low serum oestradiol and the inappropriately low gonadotrophin (LH and FSH) concentrations, with the low serum free thyroxine and inappropriately low serum TSH concentration, are suggestive of deficiency of pituitary or hypothalamic hormone secretion. Although the serum prolactin concentration is within the reference range, it should be considerably higher than this during lactation.

Comment: these results are typical of panhypopituitarism. Because she conceived normally and the symptoms only developed after delivery, it is probable that pituitary dysfunction was of recent origin. Possible causes include:

- a rapidly growing tumour;
- infarction;
- autoimmune destruction (very rare).

The most likely diagnosis was postpartum haemorrhage causing pituitary infarction (Sheehan's syndrome). The pituitary gland enlarges during pregnancy and is vulnerable to ischaemia due, for example, to hypotension following postpartum haemorrhage.

A combined pituitary stimulation test was performed to confirm the diagnosis and to determine the pituitary reserve (p.**176**). For such a test, the usual dose of insulin, to achieve adequate stress-associated hypoglycaemia, is 0.15 U/kg bodyweight (BW). However, as hypothalamic pituitary hypofunction was suspected, this dose was reduced to 0.1 U/kg BW. The results are shown below.

Combined pituitary stimulation test

Time minutes	Glucose mmol/L	Cortisol nmol/L	GH mU/L	TSH mU/L	Prolactin mU/L	FSH U/L	LH U/L
0	4.5	210	1.6	0.7	558	1.7	0.6
Intravenous insulin 0.1 U/kg BW, TRH 200 μg, GnRH 100 μg							
30	1.9	255	1.5	0.6	600	1.3	1.8
60	1.4	370	2.8	1.6	620	1.5	2.2
90	4.9	420	2.9				
120	4.2	425	3.3				

Interpretation: the patient developed symptomatic hypoglycaemia, 60 minutes after receiving insulin, and the plasma glucose fell to 1.4 mmol/L. Intravenous glucose was given and the test continued. Although the plasma cortisol concentration rose by more than 200 nmol/L, this response was impaired because it failed to exceed 550 nmol/L despite the adequate hypoglycaemia. Likewise the plasma growth hormone concentration did not rise significantly nor the plasma TSH and prolactin, and FSH and LH concentrations after the intravenous administration of TRH and GnRH respectively. These results confirmed panhypopituitarism.

Comment: during a combined pituitary stimulation test (insulin stress test) glucose must be readily available and given intravenously if the patient develops symptomatic hypoglycaemia, but it is important not to induce hyperglycaemia because this is also dangerous, especially if the rise in the extracellular glucose concentration has been rapid. The test may be continued as it is the symptoms associated with a low plasma glucose concentration that stimulate stress hormone release. At the end of the test the patient must be given something to eat.

Diagnosis: pituitary necrosis causing panhypopituitarism following a postpartum haemorrhage (Sheehan's syndrome).

CASE 6.4   SECONDARY AMENORRHOEA

A 23-year-old woman was referred for investigation of amenorrhoea. Her periods had started when she was 14 and became regular occurring every 28 days and lasting for five days. She had never been prescribed 'hormone replacement'. Eighteen months before presentation she moved to London to work as a secretary living in a flat away from home for the first time. She had lost approximately 5 kg in weight to approximately 50 kg and had become amenorrhoeic six months before presentation. On examination, she was underweight but had normal secondary sex characteristics; she was not hirsute. She was clinically euthyroid and her blood pressure was normal.

Comment: normal menses had previously been established and therefore causes of primary amenorrhoea, such as chromosomal abnormalities (Turner's syndrome; 45,X), can be excluded. The most likely diagnosis is stress-related secondary amenorrhoea associated with weight reduction. It is important to find out approximately how heavy she was at the time of menarche as this may give some indication as to the amount of weight gain anticipated before menstruation can be expected to start again. She weighed 58 kg at the menarche.

It is important to exclude hyperprolactinaemia and hyperthyroidism as causes of secondary amnorrheoa (Table 6.5). The following investigations were performed:

Serum			Reference range
LH	1.2	U/L	1.4–11.6 (follicular phase)
FSH	4.0	U/L	4.1–9.5 (follicular phase)
Prolactin	376	mU/L	<425
Total T$_4$	116	nmol/L	60–160
TSH	0.8	mU/L	0.5–2.9

Interpretation: the results of the preliminary investigations were essentially normal with the serum LH and FSH concentrations at the lower end of the reference range consistent with hypogonadotrophic hypogonadism. The serum gonadotrophin concentrations must be interpreted against a reference range determined for the same phase as the menstrual cycle; because she was not menstruating it must have been equivalent to the preovulatory (follicular) phase. There was no evidence of hyperthyroidism or hyperprolactinaemia.

Table 6.5 Some causes of secondary amenorrhoea

Physiological
    pregnancy
Primary ovarian failure
    menopause
    premature ovarian failure
Hypogonadotrophic hypogonadism
    stress, severe illness
    weight loss, malnutrition, obesity
Hypotholamic–pituitary disorders
    hyperprolactinaemia
    neoplastic, granulomatous and other infiltrative disorders
    vascular disorders such as Sheehan's syndrome
Increased sex steroid synthesis
    polycystic ovary syndrome
    congenital adrenal hyperplasia (including late onset)
    tumours of the adrenal glands or ovaries
Hyperthyroidism

Comment: if hyperprolactinaemia is suspected, it is important to examine the patient for the presence of galactorrhoea. However, a blood sample for the measurement of plasma prolactin should be taken before the examination of the breasts. Because prolactin secretion is stimulated by stress the sample must be taken with the minimum of 'trauma'; in some circumstances, it may be necessary to insert an indwelling cannula approximately 30 minutes before taking the blood sample.

Progress: a diagnosis of stress-related secondary amenorrheoa associated with weight reduction was made and the patient was encouraged to increase her body weight. She returned five months later and had put on 3.5 kg. Two months later regular menses resumed.

Diagnosis: stress-related secondary amenorrheoa associated with weight reduction.

CASE 6.5 CUSHING'S DISEASE

A 68-year-old woman with longstanding noninsulin-dependent diabetes mellitus and poorly controlled hypertension was referred to an endocrinologist. On examination, she had central obesity and was plethoric; her blood pressure was 210/115 mmHg. The diagnosis of Cushing's syndrome was considered and the following investigations were performed:

Plasma			Reference range
Creatinine	83	μmol/L	55–110
Urea	6.3	mmol/L	2.5–7.0
Sodium	143	mmol/L	135–145
Potassium	3.6	mmol/L	3.5–4.8
T_{CO_2} (bicarbonate)	33	mmol/L	24–32
Glucose	22.0	mmol/L	3.6–6.5
Cortisol (09.00 hours)	523	nmol/L	150–700

Urine
Cortisol   1245 nmol/24 hours   <250

Overnight dexamethasone suppression test (2 mg) (p.179)

Plasma
Cortisol (09.00 hours)   531   nmol/L   <100

Interpretation: there was moderately severe hyperglycaemia and, although a single result must be interpreted with caution, this was suggestive of poor control of glucose. The plasma potassium concentration was low-normal and the Tco_2 slightly raised; the plasma sodium concentration was at the upper end of the reference range. These findings are compatible with increased mineralocorticoid activity. The first plasma cortisol concentration, in the specimen taken at 09.00 hours, was within the reference range, but there is failure to suppress cortisol output following 2 mg dexamethasone; the 24-hour urinary free-cortisol excretion was significantly increased.

Comment: the initial biochemical findings were consistent with:

- *excess mineralocorticoid secretion*, causing enhanced sodium reabsorption in exchange for either potassium or hydrogen ions in the distal renal tubules resulting in a hypokalaemic alkalosis;
- *excess glucocorticoid secretion* with impaired glucose tolerance or diabetes mellitus and obesity.

Failure of the glucocorticoid drug, dexamethasone, to inhibit ACTH release, and, therefore to reduce in the plasma cortisol concentration, suggested that normal feedback regulation of cortisol production was impaired. The high 24-hour urinary free-cortisol excretion indicated that the overall rate of cortisol production was increased even though the plasma cortisol concentration at 09.00 hours was normal.

These abnormalities indicated a probable diagnosis of Cushing's syndrome and prompted further investigations to elucidate the cause.

Further investigations:

Plasma				**Reference range**
ACTH (09.00 hours)	19	ng/L		10–80

High-dose dexamethasone suppression test (2 mg six-hourly for 48 hours)
Cortisol (09.00 hours, day 1)   449   nmol/L
Cortisol (09.00 hours, day 3)   44   nmol/L

Comment: the plasma ACTH concentration is not suppressed. This excludes the diagnosis of an adrenal adenoma or carcinoma. The most likely diagnosis is a pituitary adenoma. Although typically the plasma ACTH concentration is moderately raised in this condition, it is sometimes within the reference range. The suppression of cortisol production with high-dose dexamethasone is also consistent with a diagnosis of pituitary-dependent Cushing's syndrome, although some occult ACTH secreting tumours also suppress under these conditions.

Progress: an MRI scan was performed, which demonstrated an asymmetrically enlarged pituitary gland, consistent with an adenoma. This was subsequently successfully removed by trans-sphenoidal surgery.

Diagnosis: pituitary-dependent Cushing's disease.

Thyroid disorders

7

CASE 7.1

A 72-year-old widow was admitted to the intensive care unit in January, having been found semi-comatose at home. No history was available at the time of admission.

On examination she was overweight; her face was 'puffy' and there was marked bilateral pitting ankle oedema. She responded to verbal commands and was able to move all four limbs. The blood pressure was 110/60 mmHg, slow atrial fibrillation was present and the heart sounds were faint. The respiratory rate was 16 per minute and the rectal temperature 31°C. Chest expansion was poor and there was consolidation at the right lung base. There was no clinical evidence of a focal neurological defect. The deep tendon reflexes were reduced and delayed. The abdomen was tense; bowel sounds were faint but present. The following biochemical investigations were performed:

Plasma			Reference range
Creatinine	110	μmol/L	60–110
Sodium	127	mmol/L	135–145
Potassium	4.8	mmol/L	3.5–4.8
T_{CO_2} (bicarbonate)	21	mmol/L	22–32
CK	364	U/L	<200
HBD (LD$_1$)	116	U/L	50–220
Amylase	894	U/L	15–300
Glucose (random)	11.6	mmol/L	3.5–6.5

QUESTIONS:

1. How would you interpret these biochemical results in relation to the patient's clinical presentation?
2. What further investigations would you request?

1. *Interpretation of biochemical results:* in the absence of a clinical history or focal neurological signs it is important to exclude a metabolic cause of loss of consciousness. There are a number of biochemical abnormalities but none is diagnostic.

Although the plasma amylase activity is raised, it is not high enough to be diagnostic of acute pancreatitis. There are many causes of a moderately raised plasma amylase activity (Table 10.1; p.**106**) but only when it exceeds approximately five times the upper reference limit and if other conditions have been excluded is the most likely cause acute pancreatitis. However, acute pancreatitis may occur in the presence of a normal plasma amylase activity.

The raised plasma CK activity may originate from skeletal or cardiac muscle and is not diagnostic of a myocardial infarct. It must be interpreted in relation to the clinical presentation and electrocardiographic (ECG) findings; this is discussed further on p.**129**.

The plasma glucose concentration is only slightly increased; the calculated plasma osmolarity is approximately 290 mmol/L. This excludes hyperosmolal nonketotic coma or diabetic ketoacidosis as a cause of the presenting symptoms.

The low plasma sodium concentration is suggestive of a dilutional hyponatraemia, the cause of which is not apparent from either the clinical presentation or associated biochemical findings. Before considering a diagnosis of inappropriate ADH secretion, other causes such as profound hypovolaemia, hypothyroidism or Addison's disease must be excluded (Table 3.2; p.**25**).

2. *Further investigation.* Important features of this case include:

- an elderly woman found in semi-coma during the winter;
- rectal temperature of 31°C;
- no focal neurological signs.

Many of these biochemical abnormalities could be related to either hypothermia or profound hypothyroidism. Therefore, further investigations must include thyroid function tests.

Serum		**Reference range**
Total T$_4$	<12 nmol/L	60–140
TSH	87 mU/L	0.2–4.0

Interpretation: the very low serum total T$_4$, and the high TSH, concentrations are diagnostic of primary hypothyroidism.

Management: once the diagnosis of hypothyroidism is considered, blood should be taken for thyroid function tests, but treatment, initially with low-dose L-thyroxine or tri-iodothyronine, should not be delayed until the results are available. The patient should be rewarmed gradually. Patients presenting with severe hypothyroidism may have an impaired adrenal response to stress despite normal basal plasma cortisol concentrations. Fluid restriction rather than fluid replacement may be required because of increased ADH release with excess water retention and dilutional hyponatraemia.

Progress: normal body temperature was restored over the following three days. She regained consciousness slowly and by day 10 was fully orientated. It transpired that hypothyroidism had been diagnosed and treated in the past but the patient had discontinued treatment approximately eight months before admission. She had become gradually more lethargic and had been confined to bed for one month before admission. Bronchopneumonia may have been a contributory factor.

Diagnosis: myxoedema coma.

CASE 7.2   THYROTOXICOSIS

A 61-year-old woman presented with a history of palpitations that had been diagnosed as multiple ventricular ectopic beats. She was initially treated with a β-blocker but developed bradycardia and tiredness. Subsequently she was treated with amiodarone for six months which was stopped in November. The following January she returned to the clinic complaining of tiredness, anxiety and insomnia.

On examination she was clinically euthyroid; there was no lid lag, exophthalmos or exaggerated tendon reflexes.

Thyroid function tests were requested because she had a family history of thyroid disease and had been on amiodarone, a drug which may cause either hypo- or hyperthyroidism. Because the plasma TSH concentration was 0.1 mU/L, thyroid function tests were repeated one month later. Serial changes were as follows:

Plasma	Date following presentation					Reference range
	10/1	12/2	26/5	8/8		
Total T_4	164	169	191	163	nmol/L	65–145
Free T_3	9.6	10.2	16.2	7.6	pmol/L	4.6–9.2
TSH	0.1	<0.1	<0.1	<0.1	mU/L	0.2–4.0
Treatment				carbimazole $\rightarrow$		

Interpretation: these results show progressive changes of hyperthyroidism with a rise in plasma total T_4 and free T_3 and feedback suppression of TSH release from the pituitary.

Comment: although this patient had subtle clinical features of hyperthyroidism, the initial thyroid function tests were considered borderline and were repeated. Those measured on February 12 were compatible with hyperthyroidism. The patient was recalled to the clinic but failed to attend for three months. By then she had overt clinical hyperthyroidism and was started on treatment with the antithyroid drug carbimazole. Ten weeks after starting treatment there was a significant fall in the serum free T_4 and T_3 levels but no detectable increase in the TSH concentration. This is a normal finding and is due to slow recovery of anterior pituitary function following prolonged feedback suppression of TSH production.

Diagnosis: thyrotoxicosis.

CASE 7.3 CONGENITAL THYROXINE-BINDING GLOBULIN (TBG) DEFICIENCY

A 46-year-old man was referred for investigation of weight loss. During the preceding six months he had lost approximately 10 kg in weight. His appetite had remained good and there had been no significant change in his bowel habit. There was no evidence of steatorrhoea. On examination his blood pressure was 140/80 mmHg, his pulse was regular at 88 per minute and his tendon reflexes were brisk. A diagnosis of thyrotoxicosis was made and thyroid function tests performed.

Plasma		Reference range
Total T_4	65 nmol/L	62–160
TSH	<0.1 mU/L	0.2–4.0

Interpretation: the plasma total T_4 concentration was within the reference range but the TSH concentration was suppressed.

Comment: the history and clinical findings were suggestive of thyrotoxicosis and this was supported by the suppressed plasma TSH concentration; however, the plasma total T_4 concentration was low-normal. The total thyroxine concentration in plasma consists of that bound to thyroxine-binding globulin (TBG), prealbumin and albumin and the very small free fraction. In this case, it is important to confirm the clinical impression of thyrotoxicosis by measuring the plasma free-hormone concentration as well as that of TBG.

It is also important to measure the plasma free T_3 concentration as this is frequently the first thyroid hormone to increase in early thyrotoxicosis and it may also be the only thyroid hormone raised in the rare condition T_3 toxicosis. Repeat thyroid function tests were requested.

Plasma		Reference range
Total T_4	62 nmol/L	62–160
Free T_4	46.2 pmol/L	11–25
Free T_3	26.5 pmol/L	2.6–8.0
TSH	<0.1 mU/L	0.2–4.0
TBG	3 mmol/L	7–17

Comment: these results confirm thyrotoxicosis associated with a low plasma TBG concentration. Causes of altered plasma TBG concentrations are shown in Table 7.1. The most likely cause in this patient was partial congenital TBG deficiency. This is an extremely rare X-linked disorder, which may be expressed as either complete or partial TBG deficiency.

This case, although a very rare condition, has been included because it illustrates some the difficulties that may occur in interpreting thyroid function tests when plasma TBG concentrations are significantly abnormal. The most sensitive test for diagnosing thyrotoxicosis is the identification of a suppressed plasma TSH concentration due to increased negative feedback by the raised plasma T_4 concentration. If the plasma free-hormone concentration had been assayed initially the diagnosis of thyrotoxicosis would have been immediately

Table 7.1 Some causes of altered plasma thyroxine-binding globulin (TBG) concentrations

Increased TBG	Decreased TBG
Pregnancy	Severe illness
Oestrogen treatment	Glucocorticoids
Chronic liver disease	Nephrotic syndrome
Inherited excess	Androgenic and anabolic steroids
	Inherited deficiency

apparent. However, most laboratories measure the plasma total T_4 and TSH concentrations first.

Diagnosis: partial thyroxine-binding globulin (TBG) deficiency in a male presenting with thyrotoxicosis.

CASE 7.4 GRAVES' DISEASE

A 46-year-old woman presented with a three-month history of malaise and myalgia. She complained of diarrhoea and had lost 8 kg in weight during this period in spite of having a good appetite. On examination, she had a diffusely enlarged thyroid gland which had a vascular bruit on auscultation. She had lid lag, but no exophthalmos. There was mild proximal muscle weakness.

Serum		Reference range
Free T_4	34.6 pmol/L	9.4–25.0
TSH	<0.1 mU/L	0.3–6.0
Thyroid stimulating antibodies	positive	

Interpretation: a raised serum free T_4, with a suppressed TSH, concentration is compatible with hyperthyroidism.

Comment: the undetectable serum TSH concentration is caused by increased feedback inhibition of anterior pituitary secretion by high serum thyroid hormone concentrations. This indicates that there is autonomous increased thyroid gland activity (primary hyperthyroidism). The presence of thyroid stimulating antibodies is diagnostic of Graves' disease.

Progress: she was started on treatment with carbimazole. The following results were obtained.

Serum	On presentation	one month	two months
Free T_4	34.6	15.8	13.1 pmol/L
TSH	<0.1	<0.1	4.3 mU/L

Although the serum free T_4 concentration fell significantly within the first month, there was no detectable rise in the serum TSH concentration until two months had elapsed. Following a period of prolonged suppression, the recovery of pituitary TSH release lags behind the fall in the serum T_4 concentration. Consequently the measurement of the serum T_4 concentration is a better indicator of thyroid status under these conditions.

Diagnosis: Graves' disease.

CASE 7.5   DRUG-INDUCED ABNORMALITIES OF THYROID FUNCTION TESTS

Two young women consulted their doctor complaining of tiredness. Patient 1 suffered from epilepsy and was on treatment with the anti-epileptic drug carbamazepine. She had a normal menstrual history. Patient 2 was on no medication apart from the oral contraceptive pill. Although there were no specific symptoms, the GP requested thyroid function tests to exclude hypothyroidism.

Serum	Patient 1	Patient 2	Reference range
T_4	55	230 nmol/L	70–160
TSH	1.0	1.5 mU/L	0.5–5.0

Interpretation: in each case the normal serum TSH concentration suggested that the patient was euthyroid. However, the total serum T_4 concentration was low in Patient 1 and raised in Patient 2.

Table 7.2   Some drugs which affect thyroid function tests

Drug	Total T_4	Free T_4	T_3	Remarks
Oestrogens	↑	N	↑	↑ TBG
Oral contraceptives	↑	N	↑	↑ TBG
Some radiocontrast media (e.g. ioponate)	↑	N	↓	Blocking $T_4 \rightarrow T_3$ (transient effect)
Amiodarone	↑	N or ↑	N	Blocking $T_4 \rightarrow T_3$
Propranolol	N	N	↓	Blocking $T_4 \rightarrow T_3$
Carbimazole	↓	↓	↓	Therapeutic effect
Propylthiouracil	↓	↓	↓	Therapeutic effect
Androgens	↓	N	↓	↓ TBG
Danazol	↓	N		Reduced TBG binding
Salicylates	↓	N		Reduced TBG binding
Phenytoin	↓	↓	N	Increased $T_4 \rightarrow T_3$
Carbamazepine	↓	↓	N	Increased $T_4 \rightarrow T_3$

N = normal; ↑ = increase; ↓ = decrease.

Comment: serum TSH secretion is under feedback control by T_4. Its plasma concentration provides a measure of biologically active thyroid hormone levels. In some laboratories, TSH alone is measured as the 'first line' screening test for thyroid disorders and further tests are only carried out if the result if abnormal. The normal serum TSH concentrations in both these cases strongly suggests that the patients were euthyroid, and nonthyroidal causes for the abnormal serum T_4 concentrations should be considered.

Patient 1 was taking carbamazepine, a drug that enhances the conversion of T_4 to the more active T_3. Many drugs cause abnormal thyroid function test results in euthyroid patients (Table 7.2). Other possible causes of a low serum T_4 concentration include a low plasma TBG concentration or hypopituitarism. However, the latter diagnosis was most unlikely in the presence of a normal menstrual history.

Patient 2 was taking an oral contraceptive pill, which presumably contained an oestrogen. Oestrogens increase serum total T_4 concentrations by inducing the synthesis of thyroxine-binding globulin. The serum concentration of the biologically active free T_4 is normal.

Diagnosis: drug-induced abnormalities of thyroid function in two euthyroid patients.

Disorders of calcium, phosphate and magnesium metabolism

8

CASE 8.1

A 52-year-old woman was admitted for a left total knee replacement. She had a 25-year history of osteoarthrosis of both knees with increasing pain and immobility during the past two years.

On examination she was slightly overweight. Her blood pressure was 150/80 mmHg and her pulse was regular at 78 per minute. Apart from gross limitation of movement of both knees, the remaining examination was unremarkable. Investigations were carried out on the day before the operation. Postoperatively she was confined to bed for a week and then started mobilization. Three weeks later, she had a further blood test before a general anaesthetic for removal of the plaster of Paris and manipulation.

Plasma	On admission	Week 3		Reference range
Creatinine	89	96	μmol/L	75–120
Urea	4.9	5.4	mmol/L	3.0–7.0
Sodium	137	139	mmol/L	133–143
Potassium	4.3	3.1	mmol/L	3.6–4.6
Total protein	71	73	g/L	62–80
Albumin	37	39	g/L	30–42
Calcium	2.74	3.26	mmol/L	2.15–2.55
Phosphate	0.79	0.57	mmol/L	0.60–1.40
ALP	525	612	U/L	90–250

X-ray report

Both knees show degenerative changes of osteoarthrosis, right worse than left. Upper right tibia shows changes of Paget's disease. Chest - normal.

QUESTIONS

1. Comment on these results.
2. What are the possible causes of hypercalcaemia in this patient?
3. What further investigations are required to identify the possible cause of the biochemical abnormalities?

1. *Comment:* the most striking abnormality is the increase in the plasma calcium concentration during the three weeks following surgery. On admission the plasma calcium was slightly raised with a low-normal plasma phosphate concentration and a raised alkaline phosphatase activity. Three weeks later the plasma calcium had risen to 3.26 mmol/L and the plasma phosphate was now low at 0.57 mmol/L. There had been a slight fall in the plasma potassium to 3.1 mmol/L. High plasma calcium concentrations are often associated with a low plasma potassium concentration, the exact cause of which is not fully understood.

Table 8.1 Some pathological causes of hypercalcaemia. A raised plasma total calcium concentration may be caused by an increase in the plasma albumin concentration, as in, for example, venous stasis during phlebotomy.

Hypercalcaemia with hypophosphataemia relative to GFR
    PTH production by the parathyroid glands
        primary hyperparathyroidism
        tertiary hyperparathyroidism
        familial hypocalciuric hypercalcaemia
    PTHRP production by nonparathyroid tissue

Hypercalcaemia with hyperphosphataemia relative to GFR
    vitamin D excess
    sarcoidosis
    extensive malignant deposits in bone
    immobilization, for example in Paget's disease
    rare causes
        idiopathic hypercalcaemia of infancy
        severe hyperthyroidism

2. *Causes of hypercalcaemia* are shown in Table 8.1. The most common cause in clinical practice, which is not listed in the table, is a rise in the plasma albumin concentration. The high plasma ALP activity raised the possibility of the rare combination of Paget's disease of bone and hypercalcaemia, most likely to occur after immobilization. Against this diagnosis was the significant fall in the plasma phosphate concentration. The high plasma calcium associated with a low plasma phosphate concentration, relative to GFR, is more suggestive of excess parathyroid hormone activity, due to either primary hyperparathyroidism or parathyroid hormone related protein (PTHRP) production from nonparathyroid malignant tissue. PTHRP is a peptide having a similar amino acid sequence to PTH at the biologically active end of the peptide. The gene that codes for PTHRP is widely distributed in body tissues but is normally suppressed. It may become derepressed in certain tumours, causing humoral hypercalcaemia of malignancy.

The differential diagnosis therefore includes:

- primary hyperparathyroidism;
- humoral hypercalcaemia of malignancy, for example PTHRP production;
- Paget's disease of bone with immobilization.

3. *Further investigations:* the following investigations were performed:

Plasma			Reference range
Albumin	37	g/L	35–55
Calcium	3.16	mmol/L	2.15–2.55
Phosphate	0.65	mmol/L	0.60–1.40
PTH	45	pg/L	10–50
Total T$_4$	131	nmol/L	60–140
TSH	0.7	mU/L	0.2–4.0

Electrophoresis
Serum and urine normal protein pattern

Comment: under normal circumstances when the plasma calcium concentration increases there is feedback suppression of parathyroid hormone (PTH) release with a fall in the plasma PTH concentration to low or undetectable concentrations. In this case, although the plasma PTH concentration was within the reference range it was inappropriately high for the raised plasma calcium concentration. These results confirm the diagnosis of primary hyperparathyroidism. It is, however, rarely necessary to measure the plasma PTH concentration because the diagnosis can usually be made by interpreting the preliminary investigations.

The patient also had Paget's disease of bone; undoubtedly the postoperative immobilization contributed to the rise in plasma calcium concentration. She was biochemically euthyroid; there was no evidence of myelomatosis which is very rarely associated with a raised plasma ALP activity.

Management: the patient was encouraged to increase oral fluid intake in order to correct any underlying volume depletion and was prescribed bisphosphonates; further mobilization was encouraged. Over the next week the plasma calcium concentration fell to 2.71 mmol/L. She subsequently had a parathyroidectomy and a single parathyroid adenoma was removed.

Diagnosis: primary hyperparathyroidism and Paget's disease of bone.

QUESTION:

1. What metabolic complication might occur immediately following a parathyroidectomy?

1. *Immediate postoperative complication after parathyroidectomy:* following the successful removal of a single parathyroid adenoma, the plasma calcium concentration may fall, even to below the reference range, because of long-standing suppression with atrophy of the other parathyroid glands in the presence of hypercalcaemia. In the absence of symptoms it is important that this fall in plasma calcium should not be treated as it stimulates the remaining parathyroid glands to resume normal function. If treatment is required the plasma calcium concentration should not be returned completely to normal as this would only cause negative feedback suppression and persistent hypoparathyroidism requiring long-term calcium and vitamin D supplementation.

The sequential biochemical changes following the parathyroidectomy are shown below:

Plasma	Day post-parathyroidectomy			
	0	**5**	**30**	
Calcium	3.01	1.74	2.08	mmol/L
Phosphate	0.83	1.86	1.37	mmol/L
Albumin	44	34	41	g/L
Urea	5.8	5.5	7.1	mmol/L

Comment: the biochemical changes on the fifth postoperative day are compatible with acquired hypoparathyroidism due to persistent suppression of parathyroid gland activity; the true free-ionized concentration is probably nearer normal than appears because the fall in albumin reduces the protein-bound fraction, as after most operations. By day 30, the plasma calcium concentration has returned towards normal and the plasma phosphate concentration has fallen. These results suggest return of parathyroid gland activity towards normal.

Diagnosis: primary hyperparathyroidism and Paget's disease of bone.

CASE 8.2   HUMORAL HYPERCALCAEMIA OF MALIGNANCY

A 53-year-old woman presented with back pain. Ten years previously, carcinoma of the breast had been diagnosed and was treated with a simple mastectomy and radiotherapy. Bony metastases in lumbar vertebrae were now diagnosed.

The following biochemical results were obtained:

Plasma			Reference range
Creatinine	128	μmol/L	55–110
Urea	7.5	mmol/L	2.5–7.0
Sodium	145	mmol/L	135–145
Potassium	3.1	mmol/L	3.5–4.8
T_{CO_2} (bicarbonate)	35	mmol/L	22–32
Albumin	36	g/L	27–42
Calcium	3.94	mmol/L	2.15–2.55
Phosphate	1.31	mmol/L	0.60–1.40
ALP	200	U/L	90–250

Interpretation: these results demonstrate severe hypercalcaemia. The plasma phosphate concentration is within the reference range, although not as high as might be expected for the low GFR, indicated by the raised plasma urea and creatinine concentrations. The plasma sodium concentration is at the upper end of the reference range. There is a mild hypokalaemic alkalosis. These findings are probably all secondary to hypercalcaemia.

Comment: the most likely cause of the hypercalcaemia is humoral hypercalcaemia of malignancy. Although primary hyperparathyroidism could produce similar biochemical changes, hypercalcaemia of such severity is unusual and can usually be excluded by demonstrating a suppressed plasma PTH concentration, but this is rarely necessary.

Hypercalcaemia of malignancy is usually caused by a humoral substance, such as PTHRP, but may occasionally arise because of the direct action of osteolytic metastases on bone. In the latter case, which is uncommon, the plasma phosphate concentration is usually high- normal or raised.

Hypercalcaemia may cause polyuria due to inhibition of the action of ADH on the renal tubule. This may result in intravascular water depletion with a rise in the plasma sodium concentration and causing a prerenal uraemia. A hypokalaemic alkalosis may also occur, the cause of which is unknown.

Diagnosis: humoral hypercalcaemia of malignancy due to breast carcinoma.

CASE 8.3   OSTEOMALACIA

A 74-year-old widower was referred to an orthopaedic surgeon for investigation of severe low-back pain. He was otherwise well and on no medication. Clinical examination was essentially normal and the following investigations were performed.

Plasma			Reference range
Creatinine	87	μmol/L	75–120
Urea	5.3	mmol/L	3.0–7.0
Sodium	141	mmol/L	133–143
Potassium	4.1	mmol/L	3.6–4.6
Calcium	2.20	mmol/L	2.25–2.60
Phosphate	0.49	mmol/L	0.85–1.40
Albumin	44	g/L	35–55
ALP	214	U/L	21–90

X-ray report
There is evidence of bone remodelling of the pelvis with coarsened trabeculation and some loss of bone density.
Bone scan report
Multiple areas of increased activity.

Interpretation: there is hypophosphataemia and a slightly low plasma calcium concentration. The plasma ALP activity is significantly increased and is most likely to be of bony origin because of the skeletal symptoms and because of the abnormalities of calcium and phosphate. There is no biochemical evidence of renal glomerular dysfunction.

Table 8.2 Some pathological causes of hypocalcaemia (low free-ionized calcium). A low plasma total calcium concentration may also be associated with a low plasma albumin concentration

---

Hypocalcaemia with low plasma PTH concentration
   idiopathic hypoparathyroidism
   acquired hypoparathyroidism

Hypocalcaemia with high plasma PTH concentration
   secondary hyperparathyroidism
      low plasma phosphate concentration
         vitamin D deficiency
            nutritional
            malabsorption
            anticonvulsant drugs
               phenytoin
               phenobarbitone
      high plasma phosphate concentration
         renal glomerular failure
   pseudohypoparathyroidism

Miscellanous conditions
   magnesium deficiency
   renal tubular disorders
   acute pancreatitis

Comment: causes of hypocalcaemia with hypophosphataemia are shown in Table 8.2. A low free-ionized calcium concentration reduces the negative feedback inhibition on PTH release from the parathyroid glands, resulting in increased circulating PTH levels (secondary hyperparathyroidism). This causes:

- stimulation of osteoclastic activity in bone, releasing calcium and phosphate into the extracellular fluid. Plasma concentrations of both calcium and phosphate rise;
- enhanced calcium reabsorption from the renal tubules but decreased phosphate reabsorption, causing phosphaturia. This tends to decrease the plasma phosphate concentration.

The most probable cause of the slight hypocalcaemia, giving rise to secondary hyperparathyroidism in this patient is vitamin D deficiency, causes of which include reduced dietary intake and insufficient exposure to sunlight. Other causes include malabsorption, anticonvulsant therapy, which increases vitamin D metabolism and inactivation and renal tubular dysfunction, associated with reduced activity of the active metabolite.

The clinical and radiological findings are consistent with osteomalacia, due to long-standing vitamin D deficiency. Unless there is an adequate supply of calcium and phosphate, osteoid cannot be calcified, despite increased osteoblastic proliferation, the hallmark of which is the raised plasma ALP activity.

Progress: the patient was started on vitamin D (calciferol) and calcium and phosphate supplementation although vitamin D alone may have been adequate if a normal diet is being taken. There was a considerable improvement in his symptoms and an increase in the plasma calcium and phosphate concentrations to normal; there was a subsequent fall in the plasma ALP activity.

Diagnosis: osteomalacia due to poor diet and insufficient exposure to sunlight.

CASE 8.4 SECONDARY HYPOPARATHYROIDISM

A 52-year-old man attended the out-patient clinic; he gave a history of progressive muscle weakness and cramps. Ten years previously he had had extensive surgery to his neck having had carcinoma of the larynx. As he spoke very little English, a more detailed history was unobtainable at the time.

On examination he had a feeding oesophagostomy. He had bilateral cataracts. There was bilateral proximal muscle weakness with a positive Trousseau's sign.

The following biochemical investigations were performed.

Plasma			Reference range
Urea	6.9	mmol/L	3.0–8.0
Sodium	140	mmol/L	133–143
Potassium	4.1	mmol/L	3.6–4.6
Albumin	29	g/L	35–55
Total protein	83	g/L	62–80
Calcium	1.79	mmol/L	2.15–2.55
Phosphate	2.40	mmol/L	0.60–1.40
ALP	138	U/L	90–250

Interpretation: the plasma calcium concentration was low even allowing for the low plasma albumin concentration at 29 g/L. The plasma phosphate concentration was high in the presence of apparently normal renal function. Plasma alkaline phosphatase activity was within the adult reference range.

Comment: causes of a low plasma calcium concentration are shown in Table 8.2. Causes of a raised plasma phosphate concentration in adults include:

- a reduced GFR, as in renal glomerular dysfunction;
- excessive growth hormone production;
- hypoparathyroidism.

The combination of a low plasma calcium and a high plasma phosphate concentration is suggestive of failure of the normal feedback mechanism due to deficient PTH activity and is compatible with hypoparathyroidism.

This patient had had extensive surgery to his neck for carcinoma of the larynx, with a total thyroidectomy and removal of all four parathyroid glands. Although he was on thyroid hormone replacement, he received no vitamin D or calcium supplementation for hypoparathyroidism. He developed one of the complications of persistent hypocalcaemia - bilateral cataracts.

Diagnosis: Secondary hypoparathyroidism following radical surgery to the neck.

CASE 8.5 MAGNESIUM DEFICIENCY - HYPOCALCAEMIA AND HYPOKALAEMIA

A 50-year-old woman with acute myeloid leukaemia was started on cytotoxic chemotherapy and prophylactic broad spectrum antibiotics. After one week she developed severe diarrhoea due to *Clostridrium difficile*. Four days later she began to complain of generalized paraesthesiae in her limbs. The following results were obtained:

Plasma			Reference range
Urea	2.9	mmol/L	3.0–7.0
Sodium	128	mmol/L	133–143
Potassium	2.4	mmol/L	3.6–4.6
Chloride	98	mmol/L	95–105
T_{CO_2} (bicarbonate)	15	mmol/L	24–32
Calcium	1.63	mmol/L	2.25–2.65
Albumin	33	g/L	35–55
Phosphate	0.54	mmol/L	0.60–1.40
Magnesium	0.29	mmol/L	0.71–1.00

Interpretation: there was profound hypomagmasaemia and hypocalcaemia, not due to hypoalbuminaemia and a hypokalaemic metabolic acidosis (low plasma T_{CO_2} concentration).

Comment: magnesium deficiency (Table 8.3) is most commonly due to increased renal or gastrointestinal loss. In this case, both causes were possible due to the use of potentially nephrotoxic drugs and the development of diarrhoea. Diarrhoea could have also been a contributing cause of the hypokalaemic acidosis (Table 4.2; p.**39**), but magnesiom deficiency *per se* is also associated with hypokalaemia and hypocalcaemia.

Magnesium deficiency is often associated with:

- hypocalcaemia as it may be an essential cofactor for the release and peripheral action of PTH;
- hypokalaemia as it is important for maintaining the normal intracellular distribution of potassium and for the reabsorption of potassium from the renal tubules.

Table 8.3 Some causes of hypomagnesaemia

Decreased intake or absorption
   dietary deficiency
   inappropriate intravenous feeding
   malabsorption
Increased renal losses
   drugs
      diuretics
      renal tubular toxins
   hypercalcaemia
   alcoholism
   osmotic diuresis
Increased nonrenal losses
   chronic diarrhoea and vomiting
   burns and sweating
Redistribution
   severe hyperparathyroidism
   metabolic acidosis
   insulin treatment

Progress: this patient was treated initially with a bolus of intravenous calcium, and was then started on a continuous intravenous infusion of magnesium and potassium. There was a gradual increase in the plasma calcium concentration which paralleled the rise in plasma magnesium concentration.

Plasma	Day 1	Day 4	Day 8	Day 12	
Potassium	2.4	3.6	3.5	4.4	mmol/L
Calcium	1.30	1.76	1.90	2.16	mmol/L
Albumin	33	34	33	35	g/L
Magnesium	0.29	0.45	0.61	0.96	mmol/L

Diagnosis: magnesium deficiency associated with hypocalcaemia and hypokalaemia.

Disorders of carbohydrate metabolism

9

Disorders of carbohydrate metabolism

9

CASE 9.1

A 41-year-old insulin-dependent diabetic woman was brought into the casualty department semi-conscious. She had a 48-hour history of diarrhoea and vomiting and had deliberately not taken insulin because she had been unwell and unable to eat.

On examination she was barely rousable and clinically dehydrated with reduced skin turgor and a dry mouth. Her respiration was deep and sighing with a rate of 30 per minute; there was an odour of ketones on her breath. Blood pressure was 90/60 mmHg and pulse rate was 100 per minute. There were course rales in both upper zones of her lungs. There were no focal neurological abnormalities; plantar reflexes were down-going. The following investigations were performed.

Plasma		**Reference range**
Glucose	67.2 mmol/L	2.5–5.5
TCO_2 (bicarbonate)	3 mmol/L	22–32
Potassium	5.8 mmol/L	3.5–5.0
Sodium	122 mmol/L	135–145
Urea	27.3 mmol/L	2.5–7.0
Creatinine	230 μmol/L	55–110

Urine		
Glucose	4+	
Ketones	2+	

Blood			
Haemoglobin (Hb)	14.7	g/dl	11.5–16.5
White cell count	43.7	×10⁹/L	4.0–11.0

QUESTIONS

1. Comment on the results of the biochemical investigations.
2. Is her total body potassium high, normal or low?
3. What would you expect the arterial blood gas results to show?

1. *Comment on the biochemical investigations.* There is severe hyperglycaemia, almost certainly due to insulin deficiency. At a time of stress, despite a reduction in caloric intake, insulin treatment should have been increased rather than stopped. Hyperglycaemia that exceeds the renal threshold for glucose causes an osmotic diuresis, and a reduction in total body fluid volume if, as in this case, the patient is unable to increase fluid intake.

The very low plasma $T\text{co}_2$ concentration is suggestive of a metabolic acidosis. Initially, the measurement of arterial blood pH and gases is essential to confirm and assess the severity of the acidosis. The finding of ketones in the urine indicates ketoacidosis.

Plasma urea and creatinine concentrations are both raised, the plasma urea more than the creatinine with respect to the upper limit of the reference ranges. This is consistent with a prerenal uraemia due to a reduced GFR as a consequence of intravascular volume depletion.

The low plasma sodium concentration is the result of a dilutional hyponatraemia with normal renal tubular function, due to the movement of water out of the intracellular into the extracellular compartment along an osmotic gradient created by the high extracellular glucose concentration. It is important to recognize that a dilutional hyponatraemia can occur in the presence of intravascular volume depletion.

Plasma, and therefore extracellular fluid, osmolality can be estimated by calculating the plasma osmolarity:

$$\text{Plasma osmolarity} = 2 \times ([Na^+] + [K^+]) + [\text{urea}] + [\text{glucose}] \text{ in mmol/L}$$

In this case:

Plasma	**Reference range**
osmolarity $= 2 \times (122 + 5.8) + 27.3 + 67.2$	
$= 350$ mmol/L	275–295

The fall in the plasma sodium concentration is an appropriate response to a rise in the extracellular osmolality as water moves out of the intracellular compartment along the osmotic gradient; a plasma sodium concentration within the reference range (135–145 mmol/L) under these circumstances would be inappropriate and would indicate significant water depletion.

2. *Assessment of total body potassium.* The total body potassium is almost certainly low, even though the plasma potassium concentration is high. There is increased urinary potassium loss due to the osmotic diuresis, causing total body potassium depletion and movement of potassium from cells into the extracellular compartment due to tissue hypoxia caused by reduced circulatory volume, a metabolic acidosis and insulin deficiency.

It is important to give adequate potassium supplementation during treatment, as the plasma potassium concentration may fall rapidly when tissue hypoperfusion, metabolic acidosis, and insulin deficiency are all corrected.

3. *Arterial blood gas results*

Blood			**Reference range**
pH	6.95		7.36–7.44
$P\text{co}_2$	2.06	kPa	4.3–6.4
$P\text{o}_2$	15.9	kPa	11.7–15.3
Actual bicarbonate	4	mmol/L	19–30

The low blood pH, associated with a low actual bicarbonate concentration indicates a metabolic acidosis. There is respiratory compensation as the blood $P\text{CO}_2$ is low; this supports the clinical observation of deep sighing respiration (Kussmaul respiration).

Management: the patient was rehydrated initially with 0.9 per cent saline, and then, after the plasma glucose concentration had fallen below 12 mmol/L, with five per cent dextrose. During the first 24 hours she received 7500 ml of fluid and 140 mmol potassium; she had a positive fluid balance of 3400 ml. She received a bolus of soluble insulin followed by an insulin infusion.

Serial biochemical results were recorded:

Plasma	On admission	eight hours	24 hours	48 hours	
Glucose	67.2	24.7	14.2	7.8	mmol/L
$T\text{CO}_2$ (bicarbonate)	3	13	18	28	mmol/L
Potassium	5.8	4.0	3.4	3.0	mmol/L
Sodium	122	143	146	144	mmol/L
Urea	27.3	22.6	9.4	3.5	mmol/L
Creatinine	230	128	66	55	µmol/L
Calculated osmolarity	350	341	322	305	mmol/L
Blood					
pH	6.95	7.27	7.44		
$P\text{CO}_2$	2.06	3.17	3.66		kPa
$P\text{O}_2$	15.9	9.9	10.8		kPa
Actual bicarbonate	4	14	17		mmol/L
Urine					
Glucose	4+	3+	1+		
Ketones	2+	3+	negative		

Comment: the plasma glucose concentration fell by more than 40 mmol/L during the first eight hours, due to increased cellular uptake and utilization stimulated by insulin administration, and increased filtration as the GFR rose. Although this fall was rather rapid, the plasma osmolarity fell by only about 10 mmol/L because the plasma sodium concentration increased as water moved back into the intracellular compartment along the osmotic gradient.

The plasma potassium concentration fell to below the reference range, indicating insufficient potassium replacement.

The arterial blood pH increased to near normal levels within 24 hours. Correction of the metabolic acidosis was largely due to:

- inhibition of ketone and H+ production by insulin;
- intravenous volume repletion with:
    restoration of GFR;
    improvement in tissue perfusion.

Diagnosis: diabetic ketoacidosis.

CASE 9.2   HAEMOCHROMATOSIS

A 54-year-old man presented to his general practitioner with a three-month history of polyuria and nocturia. Initial investigations confirmed that he had glycosuria. He had no significant past medical history. He drank approximately 40 units of alcohol per week.

On examination he was slightly overweight. His liver was palpable below the costal margin; there was no palmar erythema or telangiectasia. His blood pressure was 140/90 mmHg.

A random plasma glucose concentration was 8.4 mmol/L; he was asked to return the following morning to have a fasting plasma glucose concentration measured. This was 6.1 mmol/L. A formal glucose tolerance test was performed (p.**174**), the results of which are shown below.

Glucose load (tolerance) test

Time	0	60	120 minutes
Plasma			
Glucose	5.7	11.9	9.1 mmol/L
Urine			
Glucose	–	+	++
Ketones	+	–	–

Interpretation: ketonuria with no glycosuria is a normal response to a prolonged (overnight) fast. Although the plasma glucose concentration at 60 minutes was significantly raised at 11.9 mmol/L, the interpretation of a glucose tolerance test is based on the results obtained on the fasting and 120 minute samples. These, taken with the previous results of the random and fasting sample, confirm that this patient has impaired glucose tolerance and not diabetes mellitus. Most cases with such mildly impaired glucose tolerance are idiopathic in origin, although some may be associated with obesity; only a small proportion develop diabetes mellitus. However, it is important to exclude secondary causes of either impaired glucose tolerance or diabetes mellitus (Table 9.1).

On clinical examination there were no clinical features of acromegaly or Cushing's syndrome or a history of previous episodes of acute abdominal pain suggestive of recurrent acute pancreatitis or malabsorption with steatorrhoea; he was not taking any drugs. However, the diagnosis of chronic liver disease, such as haemochromatosis, was considered as a possible cause of impaired glucose

Table 9.1   Some secondary causes of impaired glucose tolerance

Insulin deficiency due to pancreatic disease
    chronic pancreatitis
    haemochromatosis
Relative insulin resistance or deficiency
    obesity
    excess growth hormone (acromegaly)
    excess glucocorticoid secretion (Cushing's syndrome)
Drugs
    thiazide diuretics
    steroids

tolerance as he drank excessive alcohol and had hepatomegaly. The following investigations were performed:

Plasma			Reference range
Liver function tests			
ALP	236	U/L	90–250
AST	55	U/L	5–50
ALT	40	U/L	5–45
Iron studies			
Iron	45	μmol/L	7–25
TIBC	48	μmol/L	49–78
% saturation	94	%	
Ferritin	4036	μg/L	10–400

Interpretation: essentially normal 'liver function' tests although the plasma AST activity was marginally raised; the iron studies demonstrated a raised plasma iron concentration with a markedly increased saturation of transferrin and a raised plasma ferritin concentration. These findings are compatible with iron overload.

Comment: a liver biopsy demonstrated increased periportal deposition of iron in hepatocytes with fibrosis. This histological appearance, together with the biochemical results and clinical presentation, were suggestive of haemochromatosis.

Idiopathic haemochromatosis is an uncommon autosomal recessive disorder in which increased intestinal absorption of iron produces large iron stores of parenchymal distribution (Fig. 9.1). It usually presents in middle age as cirrhosis of the liver, diabetes mellitus, hypogonadism and increased skin pigmentation. Factors such as multiple transfusions and alcohol may hasten the accumulation of iron and the development of cirrhosis. In the early stages of cirrhosis, there may be no abnormal biochemical findings but during a phase of active cellular destruction, plasma AST, and sometimes ALT, activities rise slightly, as in this case.

Diagnosis: haemochromatosis with impaired glucose tolerance.

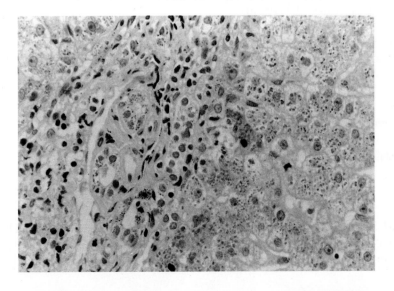

Fig. 9.1 Haemotoxylin and eosin (H and E) stain (× 400 magnification) of the liver showing numerous granules within the cytoplasm of hepatocytes and with an adjacent portal tract showing bile ducts also containing pigment within the bile duct epithelium. The severity of iron deposition is indicative of haemosiderosis and coupled with fibrosis and cirrhosis is diagnostic of haemochromatosis.

CASE 9.3   NONKETOTIC HYPEROSMOLAL COMA

A 75-year-old woman was brought into casualty in a coma. No history was available, although there was no evidence that she had been on any medication.

Plasma		Reference range
Urea	15.6 mmol/L	2.5–8.0
Sodium	156 mmol/L	135–145
Potassium	4.7 mmol/L	3.5–4.8
T_{CO_2} (bicarbonate)	20 mmol/L	22–32
Glucose	42.5 mmol/L	2.2–5.5

Urine	
Glucose	+++
Ketones	negative

Interpretation: severe hyperglycaemia and hypernatraemia with moderate uraemia. The calculated plasma osmolarity was 379 mmol/L (p.**25**). The slightly low plasma T_{CO_2} concentration is consistent with either a mild metabolic acidosis, in which case the absence of urinary ketones excludes diabetic ketoacidosis, or a respiratory alkalosis. These results are compatible with a nonketotic hyperosmolar coma.

Comment: this patient presented in nonketotic hyperglycaemic coma although she was not previously known to have diabetes mellitus.

The hypernatraemia almost certainly indicates significant water depletion. This is a result of an osmotic diuresis due to glycosuria. As the plasma glucose concentration rises, the increase in extracellular osmolality draws water out of cells, resulting in cellular dehydration and an initial fall in plasma sodium concentration due to haemodilution. Glycosuria causes an osmotic diuresis and results in total body water depletion if the patient is unable to compensate by increasing water intake. The plasma sodium concentration therefore rises. It is important to recognize that, in such circumstances, a plasma sodium concentration within the reference range is inappropriately high in the presence of hyperosmolality, and it is suggestive of water depletion.

Progress: she was rehydrated with intravenous 0.9 per cent saline and commenced on a low-dose insulin infusion. She did not regain consciousness. Six hours later she had a respiratory arrest and died.

Plasma	On admission	at six hours	
Glucose	42.5	8.6	mmol/L
T_{CO_2} (bicarbonate)	20	24	mmol/L
Potassium	4.7	4.2	mmol/L
Sodium	156	168	mmol/L
Urea	15.6	11.4	mmol/L

There had been a rapid fall in the plasma glucose concentration. The plasma sodium concentration continued to rise. This probably reflected inadequate free-water replacement, in the form of dilute saline, and a shift of water back into cells when cellular uptake of glucose was stimulated by insulin. This rapid redistribution of water is very dangerous since it may cause a rapid rise in intracranial pressure. The plasma glucose concentration should be reduced slowly in order to reduce the risk of this complication. Elderly patients in hyperosmolar nonketotic coma, due to hyperglycaemia, may be particularly sensitive to insulin.

Diagnosis: nonketotic hyperosmolar coma with fatal brain-stem compression due to coning during treatment.

CASE 9.4   REACTIVE HYPOGLYCAEMIA

A 17-year-old girl presented with a history of three recent episodes of transient loss of consciousness. All occurred in the mornings. Each had been witnessed. They were not associated with any involuntary movements or micturition but were associated with excessive perspiration.

On examination, the patient was of appropriate weight for height, blood pressure was normal and pulse 78 per minute; there were no abnormal clinical findings.

The differential diagnosis of transient loss of consciousness includes such diverse conditions as simple syncope, an inadequate glucose supply to the brain cells (neuroglycopaenia) and epilepsy. However, all three episodes occurred at approximately the same time of the day and were not associated with obvious epileptiform movements. The possibility of functional or reactive hypoglycaemia was considered and an initial random midmorning plasma glucose concentration was estimated.

Plasma		Reference range
Glucose	3.2 mmol/L	2.5–5.5

Comment: this specimen was taken in the absence of symptoms, the result was within the reference range; no further interpretation can be made. Functional or reactive hypoglycaemia can only be diagnosed by measuring serial plasma glucose concentrations following a meal or a carbohydrate load. A prolonged 75 g glucose load (tolerance) test was performed (p.**175**).

Prolonged 75 g glucose load test

Time	0	30	60	90	120	150	180	210	240	minutes
Glucose	4.6	8.7	11.7	4.2	2.3	5.3	4.6	4.1	4.6	mmol/L

Interpretation: the plasma glucose concentration fell to 2.3 mmol/L 120 minutes following the glucose load; this was associated with perspiration and a rapid bounding pulse. These results, together with the clinical findings, are compatible with reactive hypoglycaemia.

Comment: reactive hypoglycaemia is not common; the diagnosis should not be made on the basis of a low plasma glucose concentration alone. The time relation between the onset of symptoms and measured hypoglycaemia, and the relief of symptoms after glucose is given must be demonstrated.

Table 9.2 Some causes of hypoglycaemia in adults

Fasting hypoglycaemia
   insulinoma
   nonpancreatic tumours
      retroperitoneal tumours
   endocrine disorders
      pituitary failure
      adrenal insufficiency
   impaired liver function
      severe hepatitis
      liver necrosis
Nonfasting hypoglycaemia
   drugs
      insulin
      sulphonylureas
      salicylates
      alcohol
   reactive hypoglycaemia

Table 9.3 Results of plasma insulin and C-peptide estimations during hypoglycaemia (spontaneous or after a prolonged fast)

Hypoglycaemia due to:	Plasma insulin	Plasma C-peptide
Insulin administration		Low
Insulinoma or ectopic secretion	Inappropriately high	
Sulphonylurea administration		High
Alcohol		
Nonpancreatic noninsulin-secreting tumour (IGF)	Appropriately low	Low
Pituitary or adrenal failure		

Other causes of hypoglycaemia, occurring in adults, are shown in Table 9.2, the diagnosis of many depend on a detailed history and on appropriately timed specimens. For example, it is only possible to interpret a plasma insulin and C-peptide concentration if the specimens were taken during a documented hypoglycaemic episode when their concentrations would be expected to be low (Table 9.3).

Diagnosis: reactive hypoglycaemia.

CASE 9.5   MONITORING OF DIABETIC CONTROL

A 56-year-old man attended the diabetic out-patient clinic. He was on twice daily insulin injections. The results of the clinic blood and urine tests, and his own record of home blood glucose monitoring were reviewed.

Diabetic clinic results

	January	April	July	October	Reference range
Blood					
HbA$_1$	10.3	11.0	11.3	11.1 %	5.0–8.0
Urine					
Glucose	1.0	0.5	0.5	1.0 %	
Ketones	negative	negative	negative	negative	
Protein	negative	trace	trace	positive	

Home blood glucose monitoring

A typical page from his home blood glucose monitoring records using 'BM-stix':

BM-stix blood glucose (mmol/L)

September	Before breakfast	Before lunch	Before tea	Bedtime
1	7			
2		9		
3			7	
4				7
5	9			
6		7		
7			7	
8				9
9	7			
10		9		

Interpretation: the per cent blood HbA$_1$ was high at each attendance. This indicated that the mean plasma glucose concentration over the preceding six to eight weeks was also high, signifying poor diabetic control. There was glycosuria on each occasion. Assuming that the renal threshold for glucose was normal, this implied that the blood glucose concentration had exceeded approximately 11 mmol/L in the few hours immediately before each urine specimen was collected. The development of proteinuria suggests either a urinary tract infection or glomerular basement membrane dysfunction.

The home blood glucose testing results appeared to show good diabetic control. This was not consistent with the other data.

Comment: nonenzymatic glycation of haemoglobin occurs at a rate which is proportional to the plasma glucose concentration, and the per cent of blood glycated haemoglobin (HbA$_1$) is dependent on the mean plasma glucose concentration over the preceding six to eight weeks providing that the lifespan of the red blood cells is normal. In this case, the persistently high blood HbA$_1$ level suggests that overall diabetic control was poor.

Home blood glucose monitoring using test strips is a convenient method of monitoring diabetic control, especially in patients on insulin, in whom information on changes in blood glucose concentration through the day is used to adjust the dose of insulin administered. There are pitfalls to this approach, however, since some patients are unable to learn the technique, and others may deliberately falsify the results in order, for example, to avoid increases in the dose of insulin if they fear hypoglycaemic attacks. In this case, the apparent blood glucose concentrations were considerably lower than the HbA_1 results suggested. They were below the renal threshold for glucose and were less than the expected daily variation of blood glucose concentrations. This patient probably recorded false blood glucose results. Other causes of false results, such as out-dated and/or a faulty batch of test strips, were excluded.

Diagnosis: poorly controlled diabetes mellitus with false results.

Disorders of lipid metabolism

10 Disorders of lipid metabolism

CASE 10.1

A 44-year-old man was admitted to hospital with acute abdominal pain and vomiting. He had had similar, but less severe, episodes of pain during the previous six months. His bowel motions had been loose and foul smelling. He smoked approximately seven packets of cigarettes and drank about 40 units of alcohol per week.

On examination his blood pressure was 110/60 mmHg and his pulse was regular at 86 per minute. There was rebound tenderness in the epigastric region with decreased bowel sounds.

A diagnosis of acute pancreatitis was made and the following biochemical tests were requested.

Plasma			**Reference range**
Amylase	580	U/L	30–110
Urea	9.6	mmol/L	2.5–8.0
Sodium	128	mmol/L	135–145
Potassium	4.8	mmol/L	3.5–4.8
Calcium	2.06	mmol/L	2.15–2.55
Phosphate	0.68	mmol/L	0.60–1.40
Albumin	28	g/L	30–42

QUESTIONS

1 Are all these results compatible with a diagnosis of acute pancreatitis?
2 What are the likely causes of acute pancreatitis in this man?
3 What further investigations would you request, if any?

Table 10.1   Some causes of a raised plasma amylase activity

Marked increase (five to 10 times the upper reference limit)
  acute pancreatitis
  severe glomerular impairment
  severe diabetic ketoacidosis
  perforated peptic ulcer

Moderate increase (up to five times the upper reference limit)
  abdominal disorders
    perforated peptic ulcer
    acute cholecystitis
    intestinal obstruction
    abdominal trauma
    ruptured ectopic pregnancy
  salivary disorders
    mumps
    salivary calculi
    after sialography
    Sjögren's syndrome
  morphine administration
  renal glomerular dysfunction
  myocardial infarction
  acute alcohol intoxication
  diabetic ketoacidosis
  macroamylasaemia

1. *These results are compatible with but not diagnostic of acute pancreatitis* as the plasma amylase activity was only five times the upper reference limit; causes of a raised plasma amylase activity are shown in Table 10.1. The plasma calcium concentration often falls in acute pancreatitis, the cause of which is frequently attributed to precipitation of calcium soaps within the abdominal cavity. This is quantitatively unlikely; the cause can more usually be attributed to a fall in the plasma albumin concentration due to extravasation of albumin into the interstitial space as a consequence of inflammation, resulting in a fall in the albumin-bound calcium fraction. This is a common finding in any patient with shock. In this case, allowing for the fall in protein-bound calcium, the plasma free-ionized fraction was probably normal. The slightly raised plasma urea concentration was probably caused by intravascular volume depletion resulting in a prerenal uraemia.

The plasma sodium concentration was lower than might be expected in a patient who was effectively losing hypotonic fluid due to vomiting. Causes of hyponatraemia are shown in Table 3.1 (p.**24**). In this case it was important to exclude such causes as:

- pseudohyponatraemia due to hyperlipidaemia or hyperproteinaemia (p.**24**);
- dilutional hyponatraemia with a raised plasma osmolality associated with hyperglycaemia and occasionally uraemia. The high plasma glucose concentration increases the extracellular osmolality, drawing water out of the intracellular compartment and causing a dilutional hyponatraemia.

2. *Causes of acute pancreatitis* are shown in Table 10.2. The most likely cause is alcohol-related acute pancreatitis; other possible causes, such as hyperlipidaemia, must also be considered.

Table 10.2 Some causes of acute pancreatitis

Idiopathic
Biliary tract disorders
    gall stones
    obstruction
Alcohol
Metabolic
    hypercalcaemia
    hypertriglyceridaemia
Trauma or surgery
Drugs
    morphine
    steroids
Viral infections
    mumps
    infectious mononucleosis
    infectious hepatitis

3. *The following investigations should be requested:*

- plasma glucose concentration;
- examine the appearance of plasma to exclude hypertriglyceridaemia.

The following biochemical tests were performed on the initial blood sample:

Plasma		Reference range
Glucose (11.00 hours)	7.6 mmol/L	2.2–5.5
		Desirable range
Appearance	cloudy	
Total cholesterol	8.9 mmol/L	3.5–6.5
Triglyceride	17.2 mmol/L	0.5–2.2

Interpretation: the plasma glucose concentration was at the upper end of the accepted range for a nonfasting sample but not high enough to cause a significant hyponatraemia. There was combined hyperlipidaemia, the 'cloudy' appearance and significantly raised plasma triglyceride concentration almost certainly indicated a raised plasma VLDL concentration. This could cause, or be a consequence of, acute pancreatitis and could have accounted for the slightly low plasma sodium concentration (pseudohyponatraemia).

Management: this patient was treated with bed rest, intravenous fluid replacement and gastric aspiration. Computerized tomography scan of his upper abdomen revealed extensive pancreatic calcification with marked inflammation and enlargement of the head of the pancreas (Fig. 10.1). These findings were compatible with acute on chronic pancreatitis. Ten days after admission the patient was allowed home.

He was advised to stop smoking and drinking alcohol and to start a low fat/low cholesterol diet. Three months later, although he had had no further episodes of acute abdominal pain he still complained of loose, foul smelling bowel motions. A three-day faecal fat collection was requested and fasting blood lipids were measured.

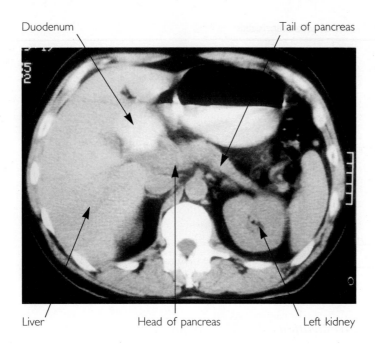

Duodenum · Tail of pancreas · Liver · Head of pancreas · Left kidney

Fig. 10.1 Computerized tomography of the upper abdomen showing extensive pancreatic calcification with marked inflammation and enlargement of the head of the pancreas. These findings were compatible with acute on chronic pancreatitis.

Plasma			**Desirable range**
Total cholesterol	8.1	mmol/L	3.4–6.5
Triglyceride	4.9	mmol/L	0.5–2.0
HDL-cholesterol	0.8	mmol/L	>0.9
LDL-cholesterol	5.2	mmol/L	2.0–4.5
			Reference range
Faecal fatty acids	32mmol/24 hours		<18

Comment: the 24-hour faecal fat result was significantly raised confirming that this patient had fat malabsorption with steatorrhoea, the most likely cause of which was chronic pancreatitis. The diagnosis of pancreatic insufficiency was confirmed by a Pancreolauryl test; he was started on pancreatic enzyme replacement.

The fasting plasma lipids confirmed a mixed hyperlipidaemia. Although they did respond to alcohol withdrawal and dietary modification, they remained significantly abnormal. He was, therefore, commenced on a fibric acid derivative; the sequential changes in the fasting plasma lipids are shown below:

Plasma	**On admission**	**three months**	**six months**	
Appearance	cloudy	cloudy	clear	
Total cholesterol	8.9	8.1	6.7	mmol/L
Triglyceride	17.2	4.9	2.4	mmol/L
HDL-cholesterol	-	0.78	1.20	mmol/L
LDL-cholesterol	-	5.2	4.2	mmol/L
Treatment		diet	diet + drug	

Diagnosis: acute pancreatitis secondary to hypertriglyceridaemia, excess alcohol ingestion and pancreatic duct outflow obstruction.

CASE 10.2 FAMILIAL HYPERCHOLESTEROLAEMIA - RESPONSE TO TREATMENT

A 26-year-old man was referred to the lipid clinic. His father had a coronary artery bypass graft at the age of 39 following a three-year history of ischaemic heart disease.

On examination he was a fit young man; he did not smoke but drank approximately 15 units of alcohol per week. There was no arcus or xanthelasma but there were tendon xanthomata on the both hands. His blood pressure was 120/70 mmHg; peripheral pulses were all present. The following biochemical tests were performed:

Plasma		Desirable range
Appearance	clear	
Total cholesterol	9.0 mmol/L	3.5–6.5
Triglyceride	1.7 mmol/L	0.5–2.2
HDL-cholesterol	1.27 mmol/L	>0.9
LDL-cholesterol	6.9 mmol/L	2.0–4.5

Interpretation: familial hypercholesterolaemia (Type IIa hyperlipidaemia).

Comment: his sister, aged 29, had a plasma cholesterol concentration of 13.0 mmol/L and his brother, aged 26, one of 11.3 mmol/L. With such a strong family history and a plasma cholesterol of 9.0 mmol/L, the patient was at high risk of developing clinical symptoms of coronary artery disease. His father should have been investigated for hypercholesterolaemia as soon as he developed symptoms at the age of 35 (Table 10.3).

Management: after eight weeks on a low fat/low cholesterol diet, the plasma total cholesterol concentration fell to 8.3 mmol/L. As he remained at significant risk of developing coronary artery disease he was started on a cholesterol-lowering drug, an HMGCoA reductase inhibitor. The plasma total-cholesterol concentration fell to 5.5 mmol/L and the LDL-cholesterol to 3.7 mmol/L.

While travelling abroad the patient stopped the tablets. On return his plasma total-cholesterol concentration had increased to 8.1 mmol/L.

Table 10.3 Recommended reasons for cholesterol screening in patients identified as being at risk of coronary heart disease

---

Personal history
    coronary or peripheral vascular disease
    associated clinical conditions
        hypertension
        diabetes mellitus
        hypothyroidism
        marked obesity
Family history (immediate relatives less than 60 years old)
    premature coronary artery disease
    hyperlipidaemia
        familial hypercholesterolaemia
        total cholesterol greater than 7.8 mmol/L

Plasma	On presentation	Diet only	Drug treatment on	off	
Total cholesterol	9.0	8.3	5.5	8.1	mmol/L
Triglyceride	1.7	1.4	1.3	0.9	mmol/L
HDL-cholesterol	1.27	1.22	1.23	1.21	mmol/L
LDL-cholesterol	6.9	6.4	3.7	6.5	mmol/L

Diagnosis: familial hypercholesterolaemia.

CASE 10.3   COMBINED HYPERTRIGLYERIDAEMIA AND HYPERCHOLESTEROLAEMIA

A 41-year-old man presented with a 10-week history of a generalized itchy papular rash that had become gradually more widespread. During the previous five weeks he had complained of progressive development of thirst, polyuria and weight loss. He drank about 32 units of alcohol per week.

On examination he was overweight. He had extensive generalized eruptive xanthomata but no arcus; his blood pressure was normal. There was marked glycosuria but no detectable ketonuria.

The following tests were performed:

Plasma		Reference range
Creatinine	65 μmol/L	50–120
Sodium	130 mmol/L	135–145
Potassium	3.7 mmol/L	3.5–4.8
T_{CO_2} (bicarbonate)	23 mmol/L	22–32
Glucose (08.30 hours)	19.5 mmol/L	2.2–5.5

		Desirable range
Appearance	turbid	
Total cholesterol	18.4 mmol/L	3.5–6.5
Triglyceride	9.9 mmol/L	0.5–2.0

Comment: the fasting plasma glucose concentration was significantly increased; this finding, together with his presenting symptoms, are diagnostic of diabetes mellitus. The low plasma sodium concentration was probably due to the raised plasma lipid concentration (pseudohyponatraemia). Many cases of hypertriglyceridaemia and mixed hyperlipidaemia are secondary to other metabolic disorders (Table 10.4).

Management: in the absence of ketonuria and a systemic metabolic acidosis he was put on a low fat diet with reduced calorie intake and started on an oral hypoglycaemic drug (metformin); alcohol intake was stopped. Within three days the plasma glucose concentration fell to less than 8.0 mmol/L; the diffuse rash disappeared within 4 weeks. Repeat plasma lipid measurements are shown below.

Table 10.4   Some causes of secondary hyperlipidaemia with associated lipid changes

Disorder	Cholesterol	Triglyceride	HDL
Alcohol excess		↑	↑
Nephrotic syndrome	↑	↑	↓
Chronic renal failure	N/↑	↑	↓
Obesity	N/↑	↑	↓
Diabetes mellitus	N/↑	↑	↓
Primary hypothyroidism	↑		
Biliary obstruction	↑		
Drugs			
beta-blockers		↑	↓
steroids		↑	
thiazides	↑	↑	↓

N = normal; ↑ = increase; ↓ = decrease.

Plasma	On presentation	Four weeks on treatment	
Appearance	turbid	clear	
Total cholesterol	18.4	6.8	mmol/L
Triglyceride	9.9	3.5	mmol/L
HDL-cholesterol	-	0.87	mmol/L
LDL-cholesterol	-	4.3	mmol/L

Diagnosis: combined hyperlipidaemia, secondary to noninsulin-dependent diabetes mellitus, excessive alcohol intake and obesity, presenting with eruptive xanthomata.

CASE 10.4   FAMILIAL HYPERCHOLESTEROLAEMIA - MYOCARDIAL INFARCT

A 50-year-old single man was admitted to the coronary care unit eight hours after the onset of central chest pain, which radiated down his left arm. He had no previous history of ischaemic heart disease. An electrocardiogram confirmed an anteriolateral myocardial infarct.

Twenty years previously he had been diagnosed as having hypercholesterolaemia. He had been prescribed clofibrate but this had been stopped after six months. He could not tolerate a low fat/low cholesterol diet and took no further treatment.

He smoked eight packets of cigarettes per week but did not drink alcohol. His father died at the age of 72 following a myocardial infarct, having had a long history of ischaemic heart disease.

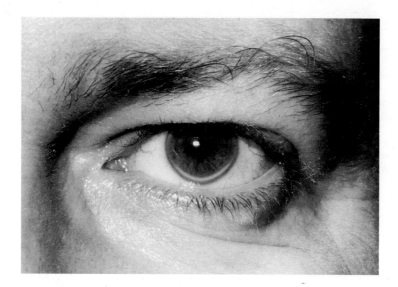

Fig. 10.2 Prominent corneal arcus in a 50-year-old man. This clinical finding is suggestive of underlying hypercholesterolaemia.

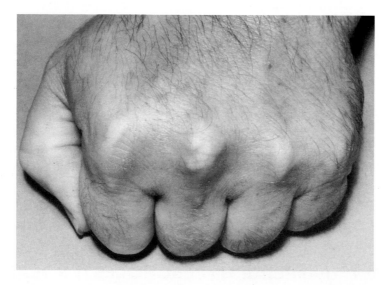

Fig. 10.3 Tendon xanthomata on the extensor tendons of the hand.

On examination he had a prominent arcus (Fig. 10.2) and tendon xanthomata on both hands (Fig. 10.3). Blood pressure was normal and peripheral pulses were all present; there was no carotid bruit. Plasma lipid measurements, performed on admission, are shown below. Thyroid function tests were normal.

Plasma			Desirable range
Appearance	clear		
Total cholesterol	12.4	mmol/L	3.5–6.5
Triglyceride	2.9	mmol/L	0.5–2.2
HDL-cholesterol	1.0	mmol/L	>0.9
LDL-cholesterol	9.8	mmol/L	2.0–4.5

Comments: following a myocardial infarct plasma lipid measurements must either be performed within 18 hours of the onset of chest pain or else the investigation should not be performed within at least three months; there is a transient fall in plasma total cholesterol concentration after myocardial infarction, the cause of which is not known.

The recommended protocol for screening for hypercholesterolaemia is shown in Table 10.3 (p.**109**). All patients should be advised to:

- stop smoking;
- take regular exercise;
- reduce the dietary saturated fat intake;
- attain ideal body weight.

Diagnosis: familial hypercholesterolaemia presenting with a myocardial infarct.

CASE 10.5   HYPERALPHALIPOPROTEINAEMIA (RAISED PLASMA HDL-CHOLESTEROL CONCENTRATION)

A 54-year-old woman was referred to the lipid clinic for advice on management of persistent hypercholesterolaemia despite appropriate dietary change. Eighteen months previously her fasting plasma lipids had been measured, at her own request, because she had become aware of the association between a high plasma cholesterol concentration and the risks of developing coronary heart disease. At that time the following results were obtained.

Plasma			Desirable range
Appearance	clear		
Total cholesterol	8.2	mmol/L	3.5–6.5
Triglyceride	1.2	mmol/L	0.5–2.2

On review in the lipid clinic she had no clinical symptoms of coronary artery disease and no family history. Physical examination was entirely normal. She was not overweight although she had lost approximately 3 kg in weight over the 18-month period; her blood pressure was 116/76 mmHg and pulse 72 per minute. There was no arcus or xanthelasma. Fasting plasma lipid measurements were repeated.

Plasma			Desirable range
Appearance	clear		
Total cholesterol	7.9	mmol/L	3.5–6.5
Triglyceride	0.9	mmol/L	0.5–2.2
HDL-cholesterol	3.56	mmol/L	>0.9
LDL-cholesterol	3.9	mmol/L	2.02–4.5

Comment: this patient had originally been diagnosed as having hypercholestero-laemia on the basis of a raised plasma total cholesterol concentration. For 18 months, she adhered to a lipid-lowering diet and was becoming anxious that her plasma cholesterol concentration remained high and that her risk of developing coronary artery disease remained unchanged. In fact, this 'patient' had an entirely normal plasma LDL-cholesterol concentration of 3.9 mmol/L but had a raised plasma HDL-cholesterol concentration, which contributed to the raised plasma total cholesterol concentration.

HDL is involved in the transport of cholesterol from cells to the liver; there is evidence that high plasma HDL concentrations reduce the risk of developing coronary artery disease. Plasma HDL-cholesterol concentrations are higher in premenopausal women than in men, and in individuals who drink alcohol exces-sively or who take exogenous oestrogens. A high plasma HDL-cholesterol concen-tration may also occur as a familial condition (hyperalphalipoproteinaemia), inherited as an autosomal dominant trait.

Although it is perfectly appropriate to screen individuals for hypercholesterol-aemia by measuring just the plasma total cholesterol concentration, if it is raised a further fasting sample should be obtained on which the total and HDL-choles-terol and triglyceride concentrations should be estimated and from which the plasma LDL-cholesterol concentration can be calculated. This calculation is only valid if the fasting plasma triglyceride concentration is less than about 5.0 mmol/L; otherwise the proportions of cholesterol and triglyceride in the VLDL particle may be abnormal, affecting the calculated result; the lipidaemia itself may interfere with the measurement of HDL-cholesterol.

Long-term management of patients with hyperlipidaemia should be based on the lipid profile, rather than on a single plasma total cholesterol concentration.

Diagnosis: familial hyperalphalipoproteinaemia with a raised plasma total choles-terol concentration.

Disorders of the liver

11

CASE 11.1

A 54-year-old woman known to have primary biliary cirrhosis was admitted to hospital with a one-week history of gradually worsening confusion and drowsiness. She had become incontinent of urine and complained of dysuria. She was on no medication. On examination she was drowsy and disorientated. When she extended her arms the typical 'liver flap' of the wrist could be demonstrated; she had palmar erythema, finger clubbing and spider naevi. She was not clinically jaundiced. The following investigations were performed.

Plasma			Reference range
Urea	4.0	mmol/L	3.0–7.0
Sodium	139	mmol/L	135–145
Potassium	1.7	mmol/L	3.6–4.6
Chloride	113	mmol/L	95–105
T_{CO_2} (bicarbonate)	12	mmol/L	24–32
Bilirubin	30	μmol/L	<17
ALP	880	U/L	90–260
AST	74	U/L	6–35

Blood			
Hb	13.6	g/dl	11.5–16.5
WBC	15.2	×10⁹/L	4.0–11.0

Urine (midstream)

Leucocytes	>100/cm³
Red blood cells	0
Culture	100 000 colonies/ml Gram-negative *Coliform bacilli*

QUESTIONS

1. Explain the results of the liver function tests in relation to the pathology of primary biliary cirrhosis.
2. Discuss the other biochemical abnormalities and state what additional biochemical investigations are indicated.

1. *Explanation of the liver function tests in terms of the pathology of primary biliary cirrhosis.* The predominant increase in plasma alkaline phosphatase (ALP) activity (approximately three times the upper adult reference limit), together with mild hyperbilirubinaemia, are typical biochemical findings of primary biliary cirrhosis and reflect intrahepatic cholestasis. Hepatocytes synthesize and release increased amounts of ALP in response to localized obstruction of bile drainage. The plasma bilirubin concentration increases due to regurgition of conjugated bilirubin into the circulation when its excretion into the bile canaliculi is impeded. The destruction of the biliary tree in primary biliary cirrhosis is usually patchy, leaving areas of normal liver which are capable of maintaining near normal bilirubin excretion. For this reason the plasma bilirubin concentration is usually only slightly raised.

Plasma aspartate aminotransferase (AST) activity is slightly increased. AST, localized in the cytoplasm and mitochondria of the hepatocyte, is released into plasma during hepatocellular necrosis, which is usually mild in primary biliary cirrhosis.

2. *Other biochemical abnormalities.* There is profound hypokalaemia, with a low plasma $T\text{co}_2$ concentration and hyperchloraemia.

Causes of hypokalaemia are shown in Table 4.2 (p.**39**). Most causes of chronic hypokalaemia are associated with a metabolic alkalosis. In this patient the low plasma $T\text{co}_2$ concentration is suggestive of a metabolic acidosis; it is too low to be due to a compensated respiratory alkalosis.

Additional biochemical investigations:

Arterial blood			**Reference range**
pH	7.28		7.36–7.44
$P\text{co}_2$	2.2	kPa	4.3–6.4
$P\text{o}_2$	11.5	kPa	11.7–15.3
Actual bicarbonate	8	mmol/L	22–30

The results confirm a metabolic acidosis with partial compensation.

In the rare case in which a reason for a metabolic acidosis is not apparent from simple tests, more information may sometimes be gained by calculating the 'anion gap', the difference between the measured cations and anions being made up of the unmeasured anions, such as protein and normally low concentrations of urate, phosphate, sulphate, lactate and other organic anions.

Reference range

Anion gap	$= ([Na^+] + [K^+]) - ([HCO_3^-] + [Cl^-])$	
	$= (139 + 1.7) - (12 + 113)$	
	$= 15.7$ mmol/L	10–17

The anion gap is normal. Consequently, conditions, such as lactic- and ketoacidosis, can be excluded from the differential diagnosis. Causes of a hyperchloraemic acidosis are shown in Table 11.1.

Comment: the most likely diagnosis is renal tubular acidosis (RTA). This diagnosis can be confirmed by measuring the urinary pH which was 7.1 in this case; a pH greater than 5.5, in the presence of a systemic metabolic acidosis, suggests failure of renal hydrogen ion excretion and supports a diagnosis of RTA.

RTA is an uncommon complication of primary biliary cirrhosis. It can be classified into two predominant types.

Table 11.1   Causes of hyperchloraemic metabolic acidosis

Renal tubular acidosis
   congenital
   acquired
Drugs
   for example carbonate dehydratase inhibitors (acetazolamide)
Transplantation of ureters into the colon, ileum or ileal loops
Profound diarrhoea

- *Classical or distal renal tubular acidosis (Type I)* is the least rare and is probably caused by altered permeability of the distal tubular cells to H^+. In milder cases this can be demonstrated by giving an ammonium chloride load (p.**171**).
- *Proximal renal tubular acidosis (Type II)* is caused by a defect in the carbonate dehydratase mechanism with impaired bicarbonate reabsorption in the proximal tubules. Associated biochemical findings include hypokalaemia, hypophosphataemia with hyperphosphaturia, glycosuria and a generalized amino aciduria.

Patients with RTA are at increased risk of developing a urinary tract infection due to the failure to acidify the urine. In this case, the urinary tract infection probably exacerbated the metabolic effects of the RTA, especially the hypokalaemia, and precipitated the hepatic encephalopathy.

Diagnosis: primary biliary cirrhosis and Type I distal renal tubular acidosis.

CASE 11.2   ACUTE HEPATITIS A INFECTION

A 43-year-old woman was admitted to hospital with a three-week history of jaundice, diffuse upper abdominal pain and generalized pruritis. Her urine had become dark and her stools were pale and smelt offensive. On examination she was jaundiced and was tender in the right hypochondrium; there was hepatomegaly.

Plasma			Reference range
Bilirubin	143	µmol/L	<17
ALP	206	U/L	21–92
AST	286	U/L	6–35

Urine	
Urobilinogen	negative
Bilirubin	++

Interpretation: increased plasma AST activity, with a relatively smaller increase in ALP activity, and hyperbilirubinaemia. The presence of bilirubinuria indicates conjugated hyperbilirubinaemia.

Comment: the increased plasma AST activity, in the presence of liver dysfunction, almost certainly reflects hepatocellular damage, causes of which include:

- hepatic hypoxia;
- viral or toxic hepatitis;
- cirrhosis of the liver, usually associated with acute hepatocellular necrosis;
- cholestasis;
- malignant or granulomatous infiltration of the liver;
- trauma or surgery to the liver.

Hepatocellular dysfunction impairs secretion of conjugated bilirubin into the biliary canaliculi, and hence into the intestine, so that the plasma conjugated bilirubin concentration increases and, being water soluble, is excreted in the urine. This results in dark urine and pale faeces. There is usually biochemical evidence of intrahepatic cholestasis, with a small increase in the plasma ALP activity.

There was no evidence of dilatation of the common or intrahepatic bile ducts on ultrasound examination.

Serum IgM antibody for hepatitis A was detected, confirming the diagnosis of infectious hepatitis.

Diagnosis: acute hepatitis A infection.

CASE 11.3   ALCOHOLIC LIVER DISEASE

A 53-year-old bachelor was brought to the accident and emergency department having been found semi-comatose. He was known to be a heavy drinker of alcohol. On examination he was jaundiced. His abdomen was distended; there was hepatomegaly and evidence of ascites. He had ankle oedema.

Plasma			Reference range
Creatinine	84	µmol/L	75–120
Urea	10.0	mmol/L	3.0–7.0
Sodium	111	mmol/L	133–143
Potassium	4.6	mmol/L	3.6–4.6
Bilirubin	166	µmol/L	<17
ALP	175	U/L	21–92
AST	371	U/L	6–35
Albumin	24	g/L	35–55
Total protein	72	g/L	62–80
Globulin	48	g/L	22–36

Interpretation: there was severe hyponatraemia with a slightly increased plasma urea concentration and a normal plasma creatinine concentration; this combination is suggestive of a prerenal uraemia.

A marked increase in plasma AST activity and bilirubin concentration with a relatively smaller increase in plasma ALP activity is suggestive of hepatocellular disease with intrahepatic cholestasis. The plasma albumin concentration was reduced but plasma total protein concentration was within the reference range, reflecting the increase in immunoglobulin concentration often present in a patient with liver disease.

Comment: although this patient had severe hyponatraemia, he had oedema and ascites and, therefore, had a higher than normal body sodium content, diluted by an even greater excess of water. Hypoalbuminaemia reduces the plasma colloid osmotic pressure, so reducing return of water from the interstitial compartment into the intravascular compartment, with the formation of oedema and contributing to the ascites. Reduced intravascular volume also results in low renal blood flow, with the development of a prerenal uraemia and stimulation of renin and aldosterone release. Aldosterone stimulates the reabsorption of sodium from the distal renal tubules. Intravascular volume depletion stimulates ADH secretion directly with enhanced free water reabsorption; a dilutional hyponatraemia may result.

The hypoalbuminaemia was probably due to decreased hepatic synthesis, although increased vascular permeability and dilution may also have been contributory factors. The increased plasma globulin concentration is probably caused by a polyclonal increase in γ globulins, which is a common finding in liver disease. In cirrhosis, there is usually an increase in plasma IgG and IgA concentrations; serum protein electrophoresis has a characteristic appearance of β–γ fusion, due to an increase in the plasma IgA concentration.

Diagnosis: decompensated alcoholic liver disease.

CASE 11.4   PRIMARY BILIARY CIRRHOSIS

A 58-year-old woman presented with a generalized excoriating rash. This had been present and progressive for approximately three years. On examination she was clinically mildly jaundiced; there was no hepatomegaly. No urobilinogen was detectable in the urine, but there was a positive reaction for bilirubin.

Plasma			Reference range
Total bilirubin	54	μmol/L	<21
ALT	36	U/L	5–50
ALP	613	U/L	90–260

Urine	
urobilinogen	negative
bilirubin	++

Interpretation: moderately elevated plasma total bilirubin concentration with a raised plasma alkaline phosphatase activity.

Comment: the results of the urine testing, together with the raised plasma ALP activity, suggest that the raised plasma bilirubin concentration was caused by a hepatobiliary disorder rather than by haemolysis with prehepatic hyperbilirubinaemia, or by hepatitis. Possible diagnoses include:

- ascending cholangitis;
- gall stones;
- primary biliary cirrhosis;
- carcinoma of the head of the pancreas.

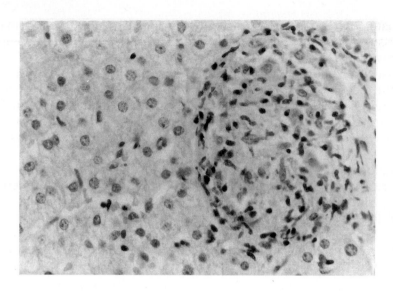

Fig. 11.1 Haematoxylin and eosin (H and E) section (× 40 magnification) of a portal tract within the liver. Normal liver is seen with an adjacent well-formed epithelioid cell granuloma which, although not discernible from the photograph, is centred in a portal tract. The finding of an epithelioid granuloma within the portal tract is highly suggestive of primary biliary cirrhosis.

Ultrasound examination of the liver and biliary tract demonstrated dilated bile ducts, but no evidence of bile stones or enlargement of the head of the pancreas. These findings, together with the clinical history, suggested that the most likely cause was primary biliary cirrhosis; further biochemical investigations to support this diagnosis include the finding of:

- mitochondrial antibodies, present in over 95 per cent of cases;
- a raised serum IgM concentration.

The diagnosis was confirmed by histological examination of a liver biopsy specimen (Fig. 11.1). In this case, the plasma ALP was derived from the hepato-biliary tract but other possible sources include osteoblasts in bone, the intestine, a rise occurring transiently following a fatty meal, or occasionally from certain tumours. If the source of the plasma ALP is not known, the individual subfractions or isoenzymes may be identified by electrophoresis.

Diagnosis: primary biliary cirrhosis.

CASE 11.5   ACUTE HEPATITIS B INFECTION

A 47-year-old woman developed intense itching and arthralgia and subsequently became jaundiced. She experienced nausea, especially after eating fatty foods. Her abdomen became distended and uncomfortable. She passed pale offensive smelling stools and dark urine. There was no previous medical history; her boyfriend was known to be a carrier of the hepatitis B virus.

On examination, she was jaundiced, had a tender, smoothly enlarged liver and moderate ascites. The following investigations were performed.

Plasma			Reference range
Total bilirubin	610	µmol/L	<21
ALP	220	U/L	21–92
AST	1820	U/L	6–35
Total protein	47	g/L	62–80
Albumin	29	g/L	35–55

Clotting studies

Prothrombin time	15.9	sec	11.5–15.0
Partial thromboplastin time	50.7	sec	25.0–37.0

Urine

Urobilinogen	increased
Bilirubin	+++

Comment: the plasma AST activity and bilirubin concentration are greatly increased; there is a proportionally smaller increase in the plasma ALP activity. This suggests hepatocellular dysfunction with secondary mild cholestasis. The fall in the total protein, mainly due to a fall in the albumin, concentration may be due to reduced hepatic protein synthesis, but could also be caused by leakage of protein out of the intravascular space through abnormally permeable vascular walls. The accumulation of interstitial fluid contributes to the ascites. The prolonged clotting times are compatible with decreased hepatic synthesis of clotting factors.

The presence of bilirubin in the urine indicates conjugated hyperbilirubinaemia, since unconjugated bilirubin circulates bound to albumin and is therefore not filtered by the glomeruli (Fig. 11.2). Urobilinogen may be detected in urine in

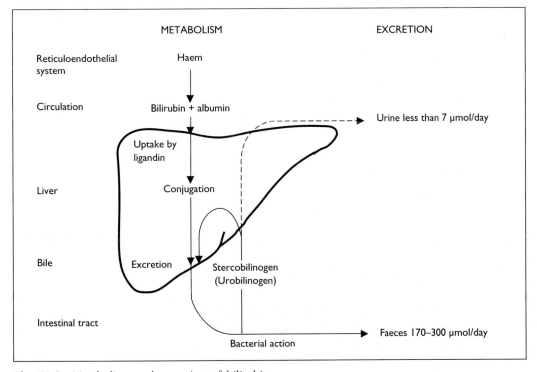

Fig. 11.2 Metabolism and excretion of bilirubin

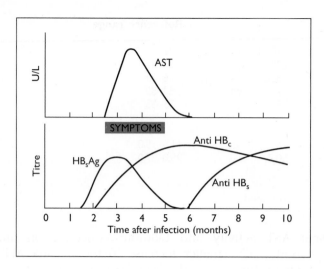

Fig. 11.3 Serological and biochemical changes following a typical infection with hepatitis B virus

healthy subjects; however, hepatic dysfunction may be associated with increased levels of excretion, due to interruption to the normal enterohepatic circulation.

Progress: she was admitted and treated with a diuretic (spironolactone) and fluid restriction to reduce the ascites. Serological tests for hepatic antibodies and serial enzymes and bilirubin assays were performed.

Plasma	On admission	Days following admission				
		3	**7**	**11**	**14**	
Total bilirubin	610	526	522	413	347	μmol/L
ALP	220	179	205	210	208	U/L
AST	1820	1688	1296	670	444	U/L

Serum
Hepatitis B surface
   antigen (HB$_s$Ag)   positive

Interpretation: there was a progressive decline in the plasma AST activity suggesting resolution of hepatocellular damage, and a fall in the plasma bilirubin concentration indicating a recovery in the capacity of the liver to secrete bilirubin. During the first two weeks, however, there was no fall in the plasma ALP activity due to persistent intrahepatic cholestasis.

In the early (prodromal) stages of the illness viral surface antigen (HB$_s$Ag) and an antigen to the internal component of the virus (HB$_e$) are present in plasma. These antigens are short-lived. During the next few weeks, an antibody response occurs with the appearance of antibodies to the viral core (HB$_c$), to HB$_e$ and finally to the surface antigen HB$_s$Ag; this may be used to document previous infection (Fig. 11.3).

Further progress: she was well enough to be discharged after 14 days. At follow-up six weeks later, she was no longer clinically jaundiced and the liver function tests were all normal.

Diagnosis: acute hepatitis B infection.

Enzymology

12

CASE 12.1

A 54-year-old man was admitted to the coronary care unit three hours after the onset of 'crushing' central chest pain. Over the previous four days he had experienced several self-limiting episodes of similar, but less severe, pain that had been precipitated by exercise. On examination he was pale and sweaty. His blood pressure was 110/90 mmHg and his pulse was regular at 78 per minute. His heart sounds were normal and his chest was clinically clear. The following investigations were requested:

Plasma			**Reference range**
Creatinine	122	μmol/L	60–120
Urea	9.2	mmol/L	2.5–8.0
Sodium	138	mmol/L	135–145
Potassium	3.2	mmol/L	3.5–4.8
CK	90	U/L	<220
HBD (LD$_1$)	174	U/L	50–220

Electrocardiogram
S–T segment elevation in leads II, III, aVF, and chest leads V2 to V6 (Fig. 12.1).

QUESTIONS

1. What is the most likely diagnosis?
2. Did the results of biochemical tests assist in making the diagnosis?
3. Which tests are important for the immediate management of a patient with this condition?

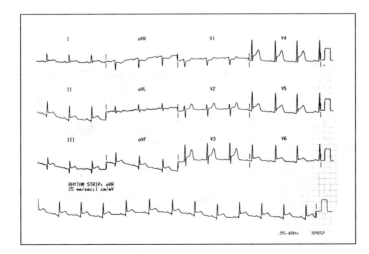

Fig. 12.1 An electrocardiographic tracing showing S–T segment elevation in leads II, III, aVF, and chest leads V2 to V6.

Table 12.1 Pattern of change of plasma cardiac enzymes after an acute myocardial infarct

Enzyme	Starts to rise (hours)	Time after infarction of peak elevation (hours)	Duration of rise (days)
CK (total)	4–6	24–48	3–5
AST	6–8	24–48	4–6
LD (HBD)	12–24	48–72	7–12

1. *The most likely diagnosis* is a myocardial infarct. Clinical presentation and electrocardiographic findings are frequently sufficient to make the diagnosis.

2. *Did these biochemical tests assist in making the diagnosis?*: the measurement of cardiac enzymes is not essential to diagnose a myocardial infarct that presents with typical clinical symptoms associated with characteristic ECG changes. Plasma creatine kinase (CK) activity starts to rise about four to six hours after the onset of chest pain. Therefore all plasma enzyme levels would be expected to be within their respective reference ranges in this patient as the blood sample was taken within four hours of the myocardial infarct. The plasma enzyme changes after an acute myocardial infarct are shown in Table 12.1.

3. *Important investigations for the immediate management of a patient following a myocardial infarct.* Plasma potassium concentration should be measured in patients suspected of having had a myocardial infarct, since both hypokalaemia and hyperkalaemia increase the risk of cardiac arrhythmia.

- *Hypokalaemia* may occur as a direct consequence of the stress associated with a myocardial infarct, when increased plasma catecholamine concentrations promote increased uptake of potassium into cells. Potassium depletion may also be present, for example in patients taking potassium-losing diuretics.
- *Hyperkalaemia* may occur as a result of renal dysfunction, this should be assessed by measuring the plasma urea and/or creatinine concentrations.

Management: the patient was admitted to the coronary care unit and given analgesia, thrombolytic therapy and oral potassium supplements. He made an uneventful recovery. The following biochemistry results were reported during the first four days of admission.

Plasma	On admission	Day 1	Day 2	Day 3	
Creatinine	122				μmol/L
Urea	9.2	7.9	5.8	7.4	mmol/L
Sodium	138	136	139	139	mmol/L
Potassium	3.2	3.8	4.2	4.2	mmol/L
CK	90	2130	714	240	U/L
AST	36	344	137	91	U/L
HBD	174	771	575	512	U/L

Comment: the rise in plasma AST and CK activities, with a peak at 24 hours, is characteristic of the changes observed after a myocardial infarct. The rise in plasma HBD activity is slightly more rapid than usual, but the slow rate of fall is typical; levels normally remain elevated for about 10 days.

In this patient the measurement of cardiac enzymes contributed little to the diagnosis of the myocardial infarct, which was made initially on clinical presentation and ECG changes. However, they may be important when the presenting features are atypical or the ECG findings are difficult to interpret due, for example, to previous ischaemic changes or to bundle branch block. Under these circumstances, the determination of more than one plasma enzyme activity (CK and HBD) on two or possibly three occasions increases the diagnostic sensitivity. If doubt remains as to the origin of a raised plasma CK activity and to the diagnosis, CK isoenzyme analysis may be helpful.

Plasma CK consists of two subunits, M and B, which combine to form three isoenzymes, BB (CK-1), MB (CK-2) and MM (CK-3). CK-MM is the predominant isoenzyme in skeletal and cardiac muscle and is detectable in the plasma of normal subjects. CK-MB accounts for about 35 per cent of the total CK activity in cardiac muscle and less than five per cent in skeletal muscle; its plasma activity is always high and usually greater than six per cent of the total activity after a myocardial infarct. CK-MB may be present in the plasma of patients with a variety of other conditions in which the total CK activity is raised but the CK-MB fraction then accounts for less than six per cent of the total.

Diagnosis: acute myocardial infarct.

CASE 12.2 CARCINOMA OF THE PROSTATE, DIAGNOSIS AND MONITORING

A 65-year-old man presented to his general practitioner with severe lower back pain. He had lost 6 kg in weight over the previous six months. His only other complaint was of urinary frequency and nocturia. On examination he was tender to palpation over the lumbar spine. He had a craggy enlarged prostate gland. An X-ray of the lumbosacral region demonstrated multiple sclerotic lesions in the lumbar spine and pelvis. The following biochemical investigations were performed:

Plasma			Reference range
ALP	457	U/L	90–250
AST	23	U/L	5–40
ALT	21	U/L	5–40
Prostate specific antigen (PSA)	58.4	µg/L	up to 2.7

Interpretation: the plasma prostate specific antigen (PSA) concentration is markedly increased; this is compatible with carcinoma of the prostate gland. The increased plasma ALP activity is most likely to be of bone origin in view of the history and the normal plasma transaminase activities.

Comment: PSA is secreted by the prostate gland. Measurement of the plasma PSA concentration is a sensitive, but nonspecific, test for diagnosing carcinoma of the prostate gland, as other disorders, such as benign prostatic hypertrophy or prostatitis, may also cause an increase in plasma PSA concentration. However, it is rare for it to exceed about 10 µg/L in these conditions. In view of this lack of specificity, the initial suspicion of carcinoma must always be followed up by other investigations such as ultrasound and biopsy.

The major sources of plasma ALP activity are bone and liver. Metastatic carcinoma could result in an increase in plasma ALP activity of either bone or liver origin. However, hepatic metastases usually result in an increase in both plasma AST and ALT activities due to destruction of surrounding hepatocytes, and in ALP activity due to cholestasis. This patient had symptoms which were probably caused by bony metastases; the sclerotic appearance on X-ray suggests that there is an osteoblastic reaction due to the infiltration of tumour cells. It is the osteoblasts that contain ALP and therefore, the increased plasma ALP activity is almost certainly of bone origin.

Management: a biopsy demonstrated a moderately differentiated prostatic carcinoma, which had spread to the local lymph nodes. He was started on treatment. This resulted in a considerable improvement in symptoms. However, about 15 months (63 weeks) later, the pain recurred. The following serial changes in plasma PSA and ALP levels were recorded:

Time	0	4	7	20	37	55	63	66	weeks
PSA	72.3	37.0	8.7	1.7	0.5	1.6	5.1	6.1	µg/L
ALP	470	421	357	194	206	195	205	200	U/L

During the first 37 weeks of treatment there was a progressive fall in the plasma PSA concentration to a low-normal concentration, accompanied by a parallel fall in plasma ALP activity. This indicated a good response to treatment. By week 55, the plasma PSA concentration had started to rise and although it was still within the reference range, it indicated that there had been further disease progression. There was no clinical evidence of relapse until week 63. The time difference between biochemical and clinical evidence of relapse is termed the 'lead time'. One of the most important uses of a tumour marker, such as PSA, is to permit early detection of recurrence of disease; few tumour markers are of value in making a diagnosis.

Diagnosis: carcinoma of the prostate with recurrence detected by a subsequent rise in plasma PSA concentration.

CASE 12.3   CHEST PAIN ? CAUSE

A 53-year-old man presented in the accident and emergency department with central chest pain; this occurred while running for a bus. During the previous six months, he had had two episodes of similar chest pain, both of which had been provoked by exercise. On this occasion an ECG, performed on admission, was entirely normal, with no evidence of ischaemic changes. However, because of his past medical history he was kept under observation. Eight hours after the onset of chest pain, plasma cardiac enzyme activities were measured. Plasma CK activity was raised at 440 U/L (reference range 0–220). He was transferred to the coronary care unit for further observation. The results of sequential cardiac enzymes are shown below:

| Plasma | Time post chest pain (hours) | | | Reference range |
	8	18	40	
CK	440	347	228 U/L	<220
HBD	180	182	179 U/L	50–220

Comment: plasma CK activity was raised on admission and showed a slow rate of fall over the next 40 hours. There was no rise in plasma HBD activity. This pattern of change is not consistent with a cardiac origin of CK and, therefore, is not compatible with a myocardial infarct. It would have been more appropriate to have done one of the following:

- repeated the plasma CK estimation approximately six hours after the first sample. If the patient had had a myocardial infarct, this would have shown a significant rise in plasma total CK activity;
- measured plasma CK isoenzyme activities. If the patient had had a myocardial infarct, the plasma CK-MB activity would have exceeded six per cent of the total enzyme activity (p.**129**).

Diagnosis: raised plasma CK activity derived from skeletal muscle.

CASE 12.4   PAGET'S DISEASE OF BONE

A 74-year-old woman presented with a six-week history of pain localized to the left buttock. It did not radiate down the lateral aspect of her thigh and was not aggravated by stooping. On examination there was full movement of both hip joints and no limitation of forward flexion or extension or lateral rotation. The left buttock was slightly warmer to touch than the right. Her blood pressure was 160/75 mmHg and pulse 84 per minute.

Plasma			Reference range
Urea	5.0	mmol/L	2.5–7.0
Albumin	39	g/L	27–42
Calcium	2.32	mmol/L	2.15–2.55
Phosphate	1.12	mmol/L	0.60–1.40
ALP	955	U/L	90–250

Radiological examination demonstrated extensive Paget's disease of the pelvis, sacrum and left femur (Fig. 12.2).

Interpretation: the only biochemical abnormality was a significant increase in the plasma alkaline phosphatase activity, compatible with the clinical and radiological findings suggestive of Paget's disease.

Comment: the clinical and biochemical findings are compatible with Paget's disease and this was supported by the radiological findings. The patient was treated with bisphosphonates for three months. There was an improvement in

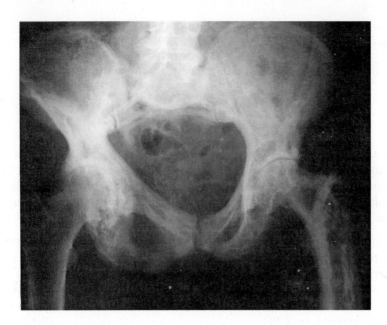

Fig. 12.2 Radiological examination demonstrating extensive Paget's disease of bone involving the pelvis, sacrum and left femur. Note the loss of the fine trabecular pattern in the shaft of the left femur compared to that in the right

symptoms with a marked relief of pain; plasma ALP activity fell to 440 U/L. Indications for treatment include:

• persistent bone pain;
• rapidly progressive bony deformities resulting in neural compression;
• high-output congestive cardiac failure;
• hypercalcaemia;
• repeated bone fractures.

In Paget's disease, there may be increased proliferation of blood vessels within the affected bone as well as cutaneous vasodilation, causing an increase in the skin temperature over the affected bones. In patients with extensive disease this increased blood flow may cause high output cardiac failure. Additional complications include pathological fractures in the involved bones. Approximately one per cent of patients develop an osteosarcoma; this is associated with a rapidly rising plasma ALP activity.

Diagnosis: Paget's disease of bone.

CASE 12.5 ALCOHOL-INDUCED CIRRHOSIS OF THE LIVER

A 75-year-old woman was admitted to hospital following the sudden onset of a right-sided hemiparesis. She had a history of epilepsy and was on treatment with phenytoin. Clinical examination confirmed a right-sided weakness; her blood pressure was normal. Her speech was slurred and her breath smelt strongly of alcohol.

Plasma			Reference range
Bilirubin	12	µmol/L	<21
ALP	275	U/L	21–92
GGT	402	U/L	6–32
AST	34	U/L	6–35
Ethanol	133	mg/dl	
Phenytoin	46	µmol/L	40–80

Interpretation: increased plasma γ-glutamyltransferase (GGT) activity, slightly increased ALP activity with a normal AST activity and bilirubin concentration; these findings are suggestive of intrahepatic cholestasis with minimal hepatocellular necrosis.

Comment: clinical signs and imaging confirmed that the hemiparesis was caused by a cerebrovascular accident (CVA). The raised plasma GGT activity could have been caused by intrinsic liver disease, such as hepatitis or by induction of hepatic enzyme synthesis by alcohol or by phenytoin and therefore must be interpreted with caution. The normal plasma AST activity is against a diagnosis of hepatitis. The associated increase in plasma ALP activity suggested that the raised plasma GGT activity was not solely due to enzyme induction as ALP synthesis is much less affected by drugs. ALP of hepatic origin may be either from hepatocytes or from the biliary tract; other causes of an increased ALP activity are shown in Table 12.2. The history of prolonged alcohol abuse was suggestive of cirrhosis and this was subsequently confirmed by histological examination of a liver biopsy specimen. The high plasma ethanol concentration was consistent with the clinical findings and added little to patient management.

Diagnosis: alcohol-induced hepatic cirrhosis, with additional induction of GGT synthesis due to continuing alcohol abuse and phenytoin medication.

Table 12.2 Some causes of a raised plasma alkaline phosphatase activity

Physiological
  normal bone growth
  pregnancy (third trimester)
  healing fractures (extensive)
Liver disease
  cholestasis (intra- or extrahepatic)
  cirrhosis
  infiltrative disorders
  space-occupying lesions (tumours, granulomas)
Bone disease
  osteomalacia and rickets
  Paget's disease of bone
  bone metastases
  extensive primary osteogenic carcinoma
  advanced primary hyperparathyroidism with bone disease
  renal osteodystrophy
Malignancy
  direct tumour production

Miscellaneous disorders

13

Miscellaneous disorders

CASE 13.1

A 68-year-old man was referred to the out-patient clinic with a history of multiple painful swellings over the joints of both hands, ankles and knees. Eight years previously he had a myocardial infarct and subsequently developed congestive cardiac failure. In the past his alcohol consumption had been high, but had been reduced to 26 units of alcohol per week (beer). His medication included:

Frusemide	80 mg	bd
Amiloride	5 mg	od
Dyazide (triamterine and hydrochlorothiazide)	1 tab	od
Captopril	50 mg	tds
Naproxen	250 mg	bd

On examination there were multiple tender subcutaneous swellings involving the interphalangeal joints of both hands, elbows, ankles and knees, and the interphalangeal joints of the second toes of both feet. Some of these had ulcerated, exposing subcutaneous white material (Fig. 13.1). There were no signs of congestive cardiac failure; examination was otherwise unremarkable. The results of initial biochemical investigations are shown:

Plasma		**Reference range**
Creatinine	210 μmol/L	60–120
Urea	26.8 mmol/L	2.5–8.0
Sodium	136 mmol/L	135–145
Potassium	4.1 mmol/L	3.5–4.8
T_{CO_2} (bicarbonate)	26 mmol/L	22–32
Urate	0.73 mmol/L	0.17–0.44

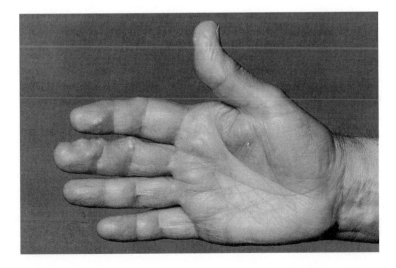

Fig. 13.1 Ulcerated lesions on the fingers of the right hand

QUESTIONS

1. Comment on these results.
2. What is the differential diagnosis?
3. What are possible causes of the biochemical abnormalities?

1. *Comment:* the plasma urea and creatinine concentrations are high, indicating renal glomerular dysfunction. This is almost certainly chronic in view of the raised plasma creatinine concentration. However, the plasma urea concentration is proportionally more elevated than the plasma creatinine concentration with respect to the upper reference limit, suggesting that there may be a degree of intravascular volume depletion resulting in acute on chronic renal glomerular dysfunction.

There is severe hyperuricaemia, which cannot be attributed to glomerular retention alone, since such a high plasma urate concentration would not be expected to occur until the GFR had fallen sufficiently to increase the plasma urea concentration to at least 40 mmol/L. This suggests that there may also be primary hyperuricaemia.

2. *The differential diagnosis:* the clinical and biochemical findings are consistent with:

- chronic tophaceous gout;
- acute gouty arthropathy;
- chronic renal glomerular dysfunction.

Table 13.1   Some causes of hyperuricaemia

Increased purine synthesis
    primary hyperuricaemia
    glucose-6-phosphatase deficiency
Increased dietary purine intake
    red meat
Increased nucleic acid turnover
    any malignancy, aggravated by chemotherapy
        leukaemia
        lymphoma
    polycythaemia rubra vera
    psoriasis
    tissue breakdown
    trauma
    acute starvation
Decreased purine recycling
    Lesch–Nyhan syndrome
Decreased renal urate excretion
    renal glomerular dysfunction
    diuretics
    salicylates (low-dose)
    ethanol
    hypercalcaemia
    lactic acidosis

---

3. *Possible causes of the biochemical abnormalities:* causes of hyperuricaemia are shown in Table 13.1. In this patient the most likely contributing factors are:

- *increased purine synthesis:*
  primary hyperuricaemia;
- *decreased renal urate excretion:*
  glomerular dysfunction, caused by:
    - drugs: combined nonsteroidal anti-inflammatory drugs (NSAID) and diuretic therapy (nephrotoxic);

- urate nephropathy (unlikely to be the only cause of glomerular failure but it may accelerate renal damage).

diuretic-induced intravascular volume contraction with reduced GFR;

thiazide diuretics;

ethanol.

Progress: naproxen (an NSAID) therapy was stopped and colchicine started, to control the acute gouty arthropathy. Diuretic therapy was reduced and he was advised to stop drinking alcohol. After one week, the pain was considerably relieved, although he now suffered slight exertional dyspnoea; there were clinical signs of mild congestive cardiac failure. The following plasma biochemical results were obtained:

Plasma	At presentation	At one week	
Creatinine	210	151	μmol/L
Urea	26.8	10.7	mmol/L
Sodium	136	143	mmol/L
Potassium	4.1	3.6	mmol/L
T_{CO_2} (bicarbonate)	26	24	mmol/L
Urate	0.73	0.54	mmol/L

Comment: the plasma urea concentration fell by more than 60 per cent, with a smaller fall in that of creatinine. These changes were probably largely attributable to an increase in extravascular volume and in renal blood flow following the reduction in diuretic therapy; they may also reflect an improvment in intrinsic renal function after stopping the NSAID therapy. There was a significant fall in plasma urate concentration, due to improvement in GFR and stopping thiazide diuretic and alcohol intake.

Diagnosis: acute on chronic gouty arthropathy precipitated by polypharmacy.

CASE 13.2   RENAL STONES – PRIMARY HYPEROXALURIA

A seven-year-old boy was referred to a paediatric urologist for investigation of recurrent abdominal pain radiating into his groin. A plain abdominal X-ray and ultrasound showed calculi in the upper and lower poles of both kidneys. There was microscopic haematuria. The following biochemical tests were performed as part of the initial investigations of the cause of renal stones.

Plasma			Reference range
Calcium	2.42	mmol/L	2.25–2.70
Phosphate	1.65	mmol/L	1.30–2.00
Albumin	43	g/L	35–55
ALP	156	U/L	70–215
Urate	0.13	mmol/L	0.06–0.24

Urine	Reference range	
Calcium	1.2 mmol/24 hour	<7.5
Oxalate	906 µmol/24 hour	70–290
Glycolate	1250 µmol/24 hour	200–790

Comment: there was significant hyperoxaluria, a normal urinary calcium excretion and normal plasma calcium and urate concentrations. Glycolate excretion was also increased. Shortly after these investigations were performed the child was admitted to hospital with partial urinary obstruction. A small, hard, pale brown, irregularly shaped stone was removed from his urethra through a cystoscope. This stone contained predominantly calcium oxalate; other important constituents of renal calculi in children, such as urate, xanthine and cystine, were excluded.

Causes of renal calculi formation are shown in Table 13.2. Although calcium oxalate stones are relatively common, this young patient had significant hyperoxaluria. Hyperoxaluria may be:

- secondary to increased oxalate intake in the diet or to increased absorption. Under normal circumstances very little oxalate is absorbed from the intestinal tract; calcium oxalate is relatively insoluble. However, in steatorrhoea, the decreased availability of calcium may allow more oxalate to be absorbed as the sodium salt. Oxalate absorption from the colon is also increased in patients who have undergone small intestinal surgery;
- primary, due to a defect in the metabolism of glyoxalate and glycolate (Fig. 13.2). In primary hyperoxaluria Type I, there is a defect in the enzyme that converts glyoxalate to glycine. Consequently glyoxalate accumulates and is metabolized by an alternative pathway to oxalate. Calcium oxalate, which has limited solubility, is excreted by the kidneys and precipitates as renal calculi. If untreated, renal dysfunction may develop.

Table 13.2 Some causes of renal calculi

Calcium-containing stones
    idiopathic
    primary hyperparathyroidism
    hyperoxaluria
        secondary
            increased dietary intake
            increased absorption
                steatorrhoea
                small intestinal resection
        primary
            hyperoxaluria Type I
    familial hypocalciuric hypercalcaemia

Mixed stones (magnesium ammonium phosphate)
    urinary infection usually with urea-splitting bacteria, high urinary pH

Urate stones
    increased urate excretion, low urinary pH

Cystine stones
    cystinuria

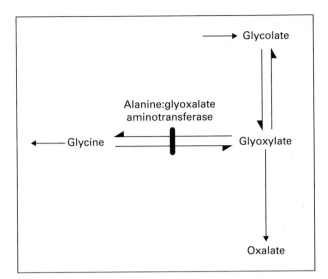

Fig. 13.2 Summary of the metabolism of glyoxylate, showing the metabolic block in Type I primary hyperoxaluria

Hyperoxaluria Type I is an uncommon condition. However, it does illustrate some important concepts:

- a possible metabolic cause of renal calculi formation should always be sought especially if recurrent and in children;
- inherited metabolic conditions may be diagnosed by demonstrating increased concentrations of precursors before the 'block' in a metabolic pathway, in this case glyoxylate;
- an inherited metabolic disorder may result in a clinical condition due to conversion of the normal enzyme substrate, by an alternative pathway to a toxic product, in this case oxalate.

Diagnosis: renal calculi caused by primary hyperoxaluria Type I.

CASE 13.3 PORPHYRIA CUTANEA TARDA

A 53-year-old woman developed a hyperpigmented blistering rash on her head and neck. Her medical history was unremarkable although she had recently been prescribed hormone replacement therapy for relief of menopausal symptoms. She drank about 21 units of alcohol per week. Because of the distribution and photosensitive nature of the skin lesions, the possibility of porphyria was considered and samples of blood, faeces and a fresh urine sample were sent to the laboratory, appropriately packed in tin foil to shield them from light. The following results were obtained.

Blood	Porphyrins	Negative
Urine	Porphobilinogen	Negative
	Porphyrins	Positive
Faeces	Porphyrins	Negative

The finding of porphyrinuria was followed up by specific porphyrin analysis on a 24-hour urine collection.

Urine		**Reference range**
Uroporphyrin	1270 nmol/24 hour	<40
Coproporphyrin	Not detected	<280

Interpretation: the screening tests were positive for urinary porphyrins; this was confirmed by demonstrating a large increase in urinary uroporphyrin excretion; coproporphyrin was not detected.

Comment: increased porphyrin or porphobilinogen excretion is characteristic of the porphyrias, a group of rare disorders, usually inherited, of haem biosynthesis. The pattern of porphyrin and/or porphobilinogen accumulation and excretion depends on the particular enzyme defect present (Fig. 13.3). The main types of porphyria are summarized in Table 13.3. The clinical presentation depends largely on which haem precursor accumulates; this depends on which enzyme is deficient.

- If porphobilinogen accumulates acute neurological disturbances, such as abdominal pain, may occur. The commonest type of acute porphyria is *acute intermittent porphyria* (AIP), in which only porphobilinogen accumulates; therefore skin lesions are absent. This usually presents as acute abdominal pain and the performance of an urgent urine screening test for porphobilinogen is indicated in order to diagnose AIP if other causes for the pain are not clinically apparent.

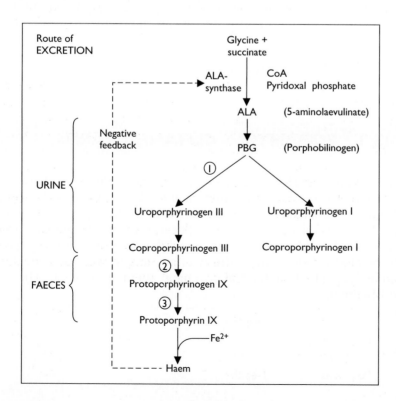

Fig. 13.3 Biosynthesis of haem, showing the sites of enzyme deficiencies in some of the porphyrias and the route by which PBG and porphyrins are excreted. 1. acute intermittent porphyria; 2. hereditary coproporphyria; 3. porphyria variegata.

Table 13.3 Clinical and biochemical features of the porphyrias

| | Hepatic porphyrias | | | | | | | Porphyrias involving the erythropoietic system | |
	Acute intermittent porphyria		Porphyria variegata		Hereditary copro-porphyria		Porphyria cutanea tarda	Congenital erythropoietic porphyria	Proto-porphyria
	Acute	Latent	Acute	Latent	Acute	Latent			
Clinical features									
Abdominal and neurological symptoms	+	–	+	–	+	–	–	–	–
Skin lesions	–	–	+	+	Rarely	Rarely	+	+	+
Biochemical abnormalities									
Urine PBG and ALA	+	+	+	–	+	–	–	–	–
Urine porphyrins	+	–	+	–	+	–	+	+	–
Faecal porphyrins	–	–	+	+	+	+	–	+	+

Acute attacks are precipitated by many drugs (for instance barbiturates, oestrogens, sulphonamides)

Symptoms may be relieved by venesection

Erythrocyte porphyrins increased

- If a porphyrin is present in excess it may cause photosensitive skin lesions. The commonest porphyria of this type is *porphyria cutanea tarda.*
- If both porphobilinogen and other porphyrins accumulate in excess, then there may be both skin and acute symptoms. This occurs in the other, rarer types of porphyrias (Table 13.3). *Lead poisoning* may cause acute abdominal symptoms; lead inhibits several of the enzymes involved in haem synthesis causing the accumulation of 5-aminolaevulinate and increased coproporphyrin excretion.

The types of samples required for the screening tests depend on which disorder is suspected clinically, as porphyrins may be excreted in either urine or faeces; in the erythropoietic porphyrias they accumulate largely in erythrocytes. Positive screening tests must be followed up by more specific analyses in order to identify the precise nature of the enzyme defect.

In this patient, the occurrence of photosensitive skin lesions and the absence of acute attacks were suggestive of porphyria cutanea tarda. This was confirmed by finding an increased urinary uroporphyrin but normal urinary coproporphyrin levels and no increased urinary porphobilinogen.

Porphyria cutanea tarda may be a dominantly inherited autosomal disorder but most cases are sporadic. The underlying genetic deficiency may be aggravated by alcohol abuse, iron overload or high-dose oestrogen therapy; the latter was probably the precipitating cause in this patient.

Diagnosis: porphyria cutanea tarda, precipitated by oestrogen therapy.

CASE 13.4 HYPOPHOSPHATAEMIC OSTEOMALACIA

A 30-year-old woman presented with a five-year history of increasing difficulty in walking. Symptoms began gradually but progressed rapidly recently. As a child she had difficulty in running and played few active sports. On examination her gait was wide-based; she had no specific neurological symptoms. There was diffuse tenderness over the thoracic and lumbar spine and pelvis. Her peripheral joints were normal. The following investigations were performed:

Plasma			Reference range
Creatinine	66	µmol/L	65–110
Urea	3.8	mmol/L	2.5–7.0
Sodium	143	mmol/L	135–145
Potassium	3.7	mmol/L	3.5–4.8
Total protein	80	g/L	60–80
Albumin	40	g/L	27–42
Calcium	2.37	mmol/L	2.15–2.55
Phosphate	0.55	mmol/L	0.60–1.40
ALP	345	U/L	90–250

Radiological examination of the lumbar spine and pelvis
The sacro-iliac joints are fused. There are a number of fractures/pseudofractures in the pelvis. Bone in the spine is abnormal. Findings compatible with metabolic bone disease or ankylosing spondylitis.

Bone scan
Extensive 'hot spots' scattered throughout the ribs, shoulder girdles, pelvis and limbs suggestive of pseudofractures.

Interpretation: low plasma phosphate concentration with slightly raised plasma ALP activity. The radiological and bone scan results are suggestive of a metabolic bone disease.

Comment: the most striking biochemical finding is the low plasma phosphate concentration in the presence of a normal plasma calcium concentration and raised ALP activity. Although it is unusual to have an unequivocally normal plasma calcium concentration, osteomalacia with secondary hyperparathyroidism is a possible diagnosis. Further investigations may help to identify the cause of the low plasma phosphate concentration. These include the measurement of renal phosphate excretion.

Plasma			Reference range
Calcium	2.30	mmol/L	2.15–2.55
Phosphate	0.51	mmol/L	0.60–1.40
Albumin	40	mmol/L	27–42
Creatinine	53	µmol/L	65–110
25-OH vitamin D	43	nmol/L	15–100
PTH	31	pmol/L	<50
Urine			
Calcium	1.65	mmol/24 hour	<7.5
Phosphate	38.4	mmol/24 hour	15–40
Creatinine	13.8	mmol/24 hour	9–17

Interpretation: the plasma phosphate concentration was low and the calcium, PTH and 25-OH vitamin D concentrations were all normal. The urinary phosphate concentration was inappropriately high for the low plasma concentration.

Comment: causes of a low plasma phosphate concentration include:

- dietary phosphate deficiency which is usually associated with a reduced urinary phosphate excretion;
- increased urinary phosphate loss due to:
  the phosphaturic effect of sustained high levels of PTH secreted in response to a fall in the plasma free-ionized calcium concentration (secondary hyperparathyroidism);
  a generalized proximal renal tubular disorder (Fanconi syndrome);
  an isolated hereditary or acquired renal tubular disorder.

In this case, the plasma calcium and PTH concentrations were normal, thus excluding the diagnosis of secondary hyperparathyroidism; there was no evidence of a generalized proximal renal tubular disorder. Therefore, the most likely diagnosis was an isolated renal tubular disorder, such as familial hypophosphataemic rickets; this condition is inherited primarily as an X-linked dominant disorder. It may present with growth retardation and rickets in childhood. Remission usually occurs following epiphyseal fusion and cessation of growth but the condition often relapses during adult life. Heterozygote adult females may present with mild hypophosphataemia and skeletal abnormalities, caused by long-standing osteomalacia.

Diagnosis: hypophosphataemic osteomalacia.

CASE 13.5 MYELOMATOSIS (MULTIPLE MYELOMA)

A 65-year-old single man, of no fixed abode, was brought into the accident and emergency department by a social worker. He gave a three-week history of a productive cough with pleuritic chest pain. He had lost approximately 7 kg in weight during the previous three months; his bowels had not opened for over one week.

On examination he was pyrexial, his blood pressure was 130/90 mmHg, his pulse 92 per minute and regular and his respiratory rate was 24 per minute; there were bilateral crackles in both lung bases. A provisional diagnosis of basal pneumonia was made and the following investigations performed.

Plasma			Reference range
Urea	4.7	mmol/L	2.5–8.0
Sodium	123	mmol/L	135–145
Potassium	3.7	mmol/L	3.5–5.0
T_{CO_2} (bicarbonate)	22	mmol/L	22–32
Total protein	124	g/L	60–80
Albumin	17	g/L	30–42
Calcium	3.22	mmol/L	2.15–2.55
Phosphate	0.93	mmol/L	0.60–1.40
ALP	224	U/L	90–250

Blood	Reference range			
Hb	8.9	g/dl		13–18
WBC	5.4	×10⁹/L		4–11
ESR	120	mm/hour		3–5

Comment: the plasma total protein concentration was significantly raised at 124 g/L and the plasma albumin concentration low at 17 g/L. If the low plasma albumin concentration is allowed for, that of free-ionized calcium was probably even higher than that suggested by the total result. The hyponatraemia was probably due to the effect of the increased protein concentration on the estimation of plasma sodium concentration (pseudohyponatraemia) (Case 3.1; p.**24**). In such cases, the sodium concentration in plasma water, and the osmolality, is normal.

The most likely diagnosis was multiple myeloma. The diagnosis depends on demonstrating:

- a monoclonal band on serum protein electrophoresis and immune paresis with or without Bence Jones proteins in urine;
- an increased proportion of atypical plasma cells in a bone-marrow aspirate;
- X-ray changes showing discrete punched out radiolucent areas, most commonly in the skull, vertebrae, ribs and pelvis. There is usually histological evidence of reduced osteoblastic activity; consequently the plasma alkaline phosphatase activity is invariably normal unless there is liver involvement.

Bone-marrow infiltration causes a normochromic, normocytic anaemia. As the plasma protein concentration increases there may be rouleaux formation of the red cells with a marked increase in the erythrocyte sedimentation rate (ESR).

All these haematological findings were present. Hypercalcaemia is a poor prognostic indicator in patients with myeloma and usually indicates widespread involvement of bone.

Further Investigations:

Serum			Reference range
Total protein	121	g/L	60–80
Paraprotein	90	g/L	
Albumin	16	g/L	30–42
IgG*	—		5.4–16.1
IgA	0.7	g/L	0.8–2.8
IgM	0.5	g/L	0.5–1.9
(*measured as paraprotein)			

Serum and urine protein electrophoresis (Fig. 13.4).
There is a monoclonal band in the γ-region (90 g/L) on serum electrophoresis, which fixes against IgG and lambda antisera; there is immune paresis. The urine protein electrophoresis shows a significant band which fixes against lambda antisera.

Comment: these findings are consistent with an IgG-lambda B-cell malignancy with Bence Jones proteinuria; myelomatosis was confirmed following examination of a bone marrow aspirate, which showed an increased number of plasma cells, many of which were atypical (Fig. 13.5).

Diagnosis: myelomatosis (multiple myeloma).

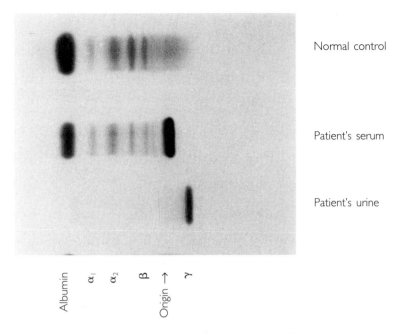

Normal control

Patient's serum

Patient's urine

Albumin α₁ α₂ β Origin → γ

Fig. 13.4 Serum and urine protein electrophoresis from a patient with myelomatosis showing a monoclonal band in the γ-region with immune paresis and a reduced albumin band and Bence-Jones protein in the urine

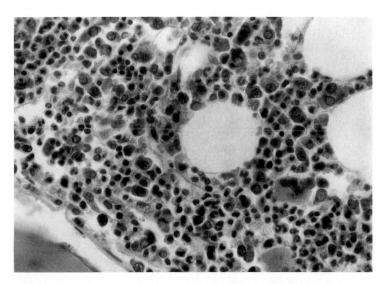

Fig. 13.5 Haematoxylin and eosin (H and E) section (× 400 magnification) shows a bone-marrow trephine with an increase in cellularity of the haemopoietic component. A variety of normal haemopoietic cells are present but, in addition, there are a number of larger cells with abundant cytoplasm, the nuclei being placed in an eccentric position. These are plasma cells and confirm the presence of multiple myeloma. Plasma cells should not exceed 20 per cent of the cells in a bone-marrow trephine

Data
interpretation,
cases and
answers

In this section, biochemical data are listed but minimal clinical information is given. Students should try to make a differential diagnosis before reading our suggested interpretation, so testing their own understanding.

Each result *must* be interpreted using the reference ranges quoted with it (p.**3**). Other data should be taken into account, because a result clearly outside the reference range may be explained if there are other abnormalities; for example, a plasma potassium concentration of about 6 mmol/L is definitely abnormal, but is expected if the plasma urea is high and there is a metabolic acidosis, suggesting renal glomerular dysfunction. Whether expected or not it may need treatment.

As in most medical situations, there is usually more than one possible explanation for the results; this can often be clarified by a logical progression of thought and, if necessary, further testing.

14.1 DISORDERS OF RENAL FUNCTION AND ELECTROLYTES

14.1.1 A 74-year-old man presented with breathlessness.

Plasma			Reference range
Urea	7.8	mmol/L	2.5–8.0
Sodium	142	mmol/L	135–145
Potassium	4.3	mmol/L	3.5–4.8
T_{CO_2} (bicarbonate)	37	mmol/L	22–32

14.1.2 An 83-year-old woman presented with dyspnoea on exertion and ankle oedema and complained of nocturia.

Plasma			Reference range
Urea	7.9	mmol/L	2.5–7.0
Sodium	128	mmol/L	135–145
Potassium	3.1	mmol/L	3.5–4.8
T_{CO_2} (bicarbonate)	36	mmol/L	22–32

14.1.3 A 64-year-old man was admitted in a confused state; he had been a heavy smoker.

Plasma			Reference range
Urea	4.3	mmol/L	2.5–8.0
Sodium	122	mmol/L	135–145
Potassium	3.7	mmol/L	3.5–4.8
T_{CO_2} (bicarbonate)	34	mmol/L	22–32

14.1.1 *Interpretation:* raised plasma $T\text{co}_2$ concentration.

Comment: the high plasma $T\text{co}_2$ concentration in a breathless patient suggests a partially or fully compensated respiratory acidosis. Since hyperkalaemia is common in acidosis, the normal plasma potassium concentration in this patient suggests that compensation is almost complete.

Diagnosis: partially compensated respiratory acidosis secondary to chronic obstructive airways disease.

14.1.2 *Interpretation:* hypokalaemic alkalosis with hyponatraemia.

Comment: the results are compatible with chronic potassium depletion and consequent metabolic alkalosis (p.**39**). Oedema indicates that the interstitial volume is increased, due to a high total body water. Despite this, the slightly raised plasma urea concentration suggests intravascular volume depletion, perhaps due to excessive use of diuretics, causing a prerenal uraemia. The mild hyponatraemia may be dilutional, due to increased free-water reabsorption by renal tubules following stimulation of ADH secretion by the low intravascular volume.

Diagnosis: treatment of congestive cardiac failure with a loop diuretic.

14.1.3 *Interpretation:* hyponatraemia with a marginally raised plasma $T\text{co}_2$ and a normal potassium concentration.

Comment: the hyponatraemia could be dilutional. The patient was confused, suggesting a rapid fall in the plasma sodium concentration; if chronic such sodium levels are well tolerated. Because of the history of heavy smoking urinary osmolalities should be measured in case there is inappropriate ADH secretion due to carcinoma of the bronchus. The slightly raised plasma $T\text{co}_2$ concentration may reflect a partially or fully compensated respiratory acidosis or a metabolic alkalosis secondary to potassium depletion.

Diagnosis: inappropriate ADH secretion (p.**25**) secondary to carcinoma of the bronchus in a patient with chronic obstructive airways disease.

14.1.4 A 48-year-old woman was diagnosed as having mild hypertension following a 'routine' medical examination.

Plasma			Reference range
Urea	5.7	mmol/L	2.5–7.0
Sodium	143	mmol/L	135–145
Potassium	3.1	mmol/L	3.5–4.8
T_{CO_2} (bicarbonate)	36	mmol/L	22–32

14.1.5 A 73-year-old man was admitted to hospital for investigation of profuse diarrhoea.

Plasma			Reference range
Urea	36.8	mmol/L	2.5–8.0
Potassium	5.4	mmol/L	3.5–4.8
Sodium	144	mmol/L	135–145
T_{CO_2} (bicarbonate)	22	mmol/L	22–32

Urine		
Sodium	18	mmol/L

14.1.6 A 52-year-old woman presented with fatigue and polyuria.

Plasma			Reference range
Creatinine	478	μmol/L	55–110
Urea	34.6	mmol/L	2.5–7.0
Sodium	132	mmol/L	135–145
Potassium	5.3	mmol/L	3.5–4.8
T_{CO_2} (bicarbonate)	14	mmol/L	22–32

14.1.7 A 71-year-old man presented with weight loss and melaena.

Plasma			Reference range
Creatinine	109	μmol/L	60–120
Urea	18.9	mmol/L	2.5–8.0
Sodium	142	mmol/L	135–145
Potassium	4.7	mmol/L	3.5–4.8
T_{CO_2} (bicarbonate)	19	mmol/L	22–32

14.1.8 A 47-year-old woman had a cholecystectomy. Two days later the following results were obtained.

Plasma			Reference range
Urea	3.2	mmol/L	2.5–7.0
Sodium	122	mmol/L	135–145
Potassium	3.9	mmol/L	3.5–4.8
T_{CO_2} (bicarbonate)	24	mmol/L	22–32

14.1.4 *Interpretation:* hypokalaemic alkalosis with a high-'normal' plasma sodium concentration.

Comment: the high-'normal' plasma sodium concentration with a low plasma potassium concentration is suggestive of excessive mineralocorticoid activity. The most likely cause of secondary hypertension (p.**46**) in this patient is either Conn's syndrome or Cushing's disease, although the patient did not have any of the clinical features of Cushing's disease.

Diagnosis: primary hyperaldersteronism (Conn's syndrome).

14.1.5 *Comment:* the plasma urea concentration is about four times the upper adult reference limit, and interpreted with the high-'normal' sodium concentration is compatible with a prerenal uraemia due to hypotonic fluid loss. The slightly raised plasma potassium and low-normal T_{CO_2} concentrations are consistent with this diagnosis. The relatively low urinary sodium concentration in the presence of intravascular volume depletion is caused by secondary hyperaldosteronism with enhanced sodium reabsorption from the distal convoluted tubules of the kidneys.

Diagnosis: prerenal uraemia secondary to intravascular hypovolaemia caused by profuse diarrhoea.

14.1.6 *Comment:* the plasma urea and creatinine concentrations are both approximately four times the upper adult reference limits and are associated with a hyperkalaemic acidosis (low plasma T_{CO_2} concentration); these results are compatible with a diagnosis of chronic renal glomerular dysfunction.

Diagnosis: chronic renal glomerular dysfunction (failure).

14.1.7 *Comment:* the plasma urea concentration is significantly increased but that of creatinine is normal. Increased breakdown of haemoglobin within the intestinal tract may increase the hepatic nitrogen load and therefore urea formation (p.**36**); in the presence of renal glomerular impairment, this may cause mild uraemia.

Diagnosis: prerenal uraemia following a gastrointestinal haemorrhage.

14.1.8 *Interpretation:* hyponatraemia with a low-normal plasma urea concentration.

Comment: these results are compatible with a dilutional hyponatraemia due to fluid overload with dextrose-saline in the immediate postoperative period. This may be compounded by increased ADH secretion in response to stress, pain or as a consequence of anaesthetic drugs (p.**25**).

Diagnosis: postoperative dilutional hyponatraemia.

14.1.9 A 28-year-old woman was admitted to hospital with a three-day history of worsening headache and photophobia; she lost consciousness and subsequently developed polyuria.

Plasma			Reference range
Urea	9.4	mmol/L	2.5–7.0
Potassium	4.0	mmol/L	3.5–4.8
Sodium	160	mmol/L	135–145
T_{CO_2} (bicarbonate)	21	mmol/L	22–32
Glucose	3.6	mmol/L	3.5–7.0

Urine		
Osmolality	94	mmol/kg

14.1.10 A 58-year-old man presented with severe shoulder tip pain.

Plasma			Reference range
Urea	5.1	mmol/L	2.5–8.0
Sodium	116	mmol/L	135–145
Potassium	3.4	mmol/L	3.5–4.8
T_{CO_2} (bicarbonate)	21	mmol/L	22–32
Osmolality	288	mmol/kg	275–295

14.1.11 A 20-year-old woman was admitted with severe emaciation.

Plasma			Reference range
Urea	10.1	mmol/L	2.5–8.0
Sodium	137	mmol/L	135–145
Potassium	1.4	mmol/L	3.5–4.8
T_{CO_2} (bicarbonate)	50	mmol/L	22–32

14.1.12 A five-week-old boy was admitted to hospital with a 10-day history of vomiting.

Plasma			Reference range
Urea	10.2	mmol/L	1.4–5.4
Sodium	135	mmol/L	127–142
Potassium	3.1	mmol/L	3.6–4.8
Chloride	75	mmol/L	95–105
T_{CO_2} (bicarbonate)	48	mmol/L	22–32

14.1.9 *Interpretation:* marked hypernatraemia with mild uraemia.

Comment: the combination of hypernatraemia and uraemia is almost certainly due to severe free-water depletion. The most common causes are either reduced water intake or increased free-water loss. The urinary osmolality is inappropriately low for the raised plasma osmolarity (341 mmol/L) and therefore indicates excessive free-water loss through the kidneys.

Diagnosis: cranial diabetes insipidus as a consequence of *Pneumococcal meningitis.*

14.1.10 *Interpretation:* marked hyponatraemia with a normal plasma osmolality.

Comment: the normal plasma osmolality indicates that, unless a significant concentration of unmeasured solute is present, the sodium activity in plasma water is probably normal. The low plasma sodium concentration is almost certainly a consequence of an increased proportion of large water-excluding molecules, such as lipid or protein (p.**24**). The plasma should have been inspected for turbidity due to lipaemia and the plasma total protein concentration measured with, if indicated, electrophoresis.

Diagnosis: myelomatosis with pseudohyponatraemia.

14.1.11 *Interpretation:* marked hypokalaemic alkalosis with mild uraemia.

Comment: the pronounced hypokalaemic alkalosis suggests total body potassium depletion but does not identify the cause (p.**39**). The mild uraemia may be caused by renal glomerular dysfunction or may indicate either a degree of intravascular volume depletion or increased protein breakdown due to severe malnourishment or a hypercatabolic state.

Diagnosis: malnutrition caused by anorexia nervosa.

14.1.12 *Interpretation:* hypochloraemic alkalosis with mild hypokalaemia and significant uraemia for age.

Comment: hypochloraemic alkalosis of this degree is almost diagnostic of vomiting due to prolonged and severe pyloric stenosis; only when there is outflow obstruction from the stomach is H^+Cl^- lost without bicarbonate. This stimulates increased gastric hydrochloric acid synthesis, which is accompanied by increased passage of bicarbonate into the extracellular fluid. The alkalosis, with potassium loss in the vomit, may result in profound hypokalaemia. Vomiting may also cause intravascular fluid volume depletion, and therefore a prerenal uraemia, with secondary hyperaldosteronism which compounds the hypokalaemic alkalosis. These findings are uncommon unless the pyloric stenosis is severe, prolonged and untreated, when fluid depletion may be sufficient to impair renal correction of the disturbance.

Diagnosis: severe and prolonged pyloric stenosis.

14.2 DISORDERS OF BLOOD GASES AND HYDROGEN ION HOMEOSTASIS

14.2.1 A 75-year-old man presented with dyspnoea and chronic cough.

Blood			Reference range
pH	7.30		7.35–7.45
P_{CO_2}	8.3	kPa	4.6–6.0
P_{O_2}	6.3	kPa	12–16
Actual bicarbonate	34	mmol/L	22–30

14.2.2 A 12-year-old girl presented with acute breathlessness.

Blood			Reference range
pH	7.10		7.35–7.45
P_{CO_2}	12.2	kPa	4.6–6.0
P_{O_2}	10.1	kPa	12–16
Actual bicarbonate	22	mmol/L	22–30

14.2.3 A blood sample was taken from a 68-year-old man immediately after a cardiac arrest.

Blood			Reference range
pH	7.13		7.35–7.45
P_{CO_2}	8.7	kPa	4.6–6.0
P_{O_2}	70.2	kPa	12–16
Actual bicarbonate	6.0	mmol/L	22–30

14.2.4 A 58-year-old man presented with oliguria and end stage renal disease.

Blood			Reference range
pH	7.12		7.35–7.45
P_{CO_2}	3.8	kPa	4.6–6.0
P_{O_2}	16	kPa	12–16
Actual bicarbonate	10	mmol/L	22–30

14.2.5 A 76-year-old man presented with an acute exacerbation of chronic obstructive airways disease.

Blood			Reference range
pH	7.36		7.35–7.45
P_{CO_2}	6.8	kPa	4.6–6.0
P_{O_2}	13.1	kPa	12–16
Actual bicarbonate	38	mmol/L	22–30

14.2.1 *Comment:* the reduced blood pH and the raised P_{CO_2} and actual bicarbonate concentration is suggestive of a partially compensated respiratory acidosis.

Diagnosis: partially compensated respiratory acidosis with a hypoxia secondary to an acute exacerbation of chronic obstructive airways disease.

14.2.2 *Comment:* the very high P_{CO_2} and the low blood pH indicate a severe respiratory acidosis due to CO_2 retention. Although the blood actual bicarbonate concentration is at the lower end of the reference range it is not sufficiently reduced to account for this degree of acidosis in a primary metabolic acidosis. The raised blood P_{CO_2} in the presence of a normal actual bicarbonate concentration is compatible with an uncompensated respiratory acidosis.

Diagnosis: uncompensated respiratory acidosis, secondary to acute reversible airways obstruction (asthma).

14.2.3 *Comment:* the blood pH is reduced. The blood P_{CO_2} is raised and the actual bicarbonate concentration markedly reduced, suggesting a combined metabolic and respiratory acidosis. The significantly raised P_{O_2} of 70.2 kPa suggests that this patient is on 100 per cent oxygen.

Diagnosis: a combined metabolic and respiratory acidosis following a cardiac arrest.

14.2.4 *Comment:* the decreased blood pH and actual bicarbonate concentration are compatible with metabolic acidosis. The blood P_{CO_2} is also decreased indicating the presence of compensatory mechanisms, with hyperventilation and increased expiration of carbon dioxide. The measurement of blood gases is rarely indicated in the management of a patient with a metabolic acidosis.

Diagnosis: a partially compensated metabolic acidosis, secondary to end-stage renal glomerular dysfunction (failure).

14.2.5 *Comment:* the low-normal blood pH associated with a raised P_{CO_2} is compatible with a respiratory acidosis. The blood actual bicarbonate concentration is increased at 38 mmol/L indicating metabolic compensation.

Diagnosis: a fully compensated respiratory acidosis.

14.2.6 The following results were obtained from an arterial blood sample, taken from a 25-year-old man in the intensive care unit who was being hyperventilated following a head injury.

Blood			Reference range
pH	7.50		7.35–7.45
P_{CO_2}	2.1	kPa	4.6–6.0
P_{O_2}	32	kPa	12–16
Actual bicarbonate	15	mmol/L	22–30

14.2.7 A five-week-old baby boy was admitted to hospital with severe projectile vomiting.

Blood			Reference range
pH	7.56		7.35–7.45
P_{CO_2}	3.4	kPa	4.6–6.0
P_{O_2}	7.6	kPa	12–16
Actual bicarbonate	56	mmol/L	22–30

14.2.8 A 43-year-old man was receiving positive pressure ventilation on a respirator following a road traffic accident.

Blood			Reference range
pH	7.54		7.35–7.45
P_{CO_2}	3.7	kPa	4.6–6.0
P_{O_2}	9.1	kPa	12–16
Actual bicarbonate	18	mmol/L	22–30

14.2.9 A 22-year-old man was admitted to hospital, having been found in a hostel in a semi-comatose condition.

Blood			Reference range
pH	7.00		7.35–7.45
P_{CO_2}	1.8	kPa	4.6–6.0
P_{O_2}	18.2	kPa	12–16
Actual bicarbonate	3.4	mmol/L	22–30

14.2.10 A 42-year-old woman was admitted to hospital with a pyrexia and nausea; she was found to have a heavy urinary tract infection.

Blood			Reference range
pH	7.20		7.35–7.45
P_{CO_2}	3.6	kPa	4.6–6.0
P_{O_2}	12.7	kPa	12–16
Actual bicarbonate	11	mmol/L	22–30

Urine

pH	5.8

14.2.6 *Comment:* the decreased blood P_{CO_2} and actual bicarbonate concentration, in the presence of a raised pH, indicates a partially compensated respiratory alkalosis. The blood P_{O_2} is greater than that which would be expected if the patient was breathing room air because the patient was receiving supplemental oxygen.

Diagnosis: a partially compensated respiratory alkalosis.

14.2.7 *Comment:* the blood actual bicarbonate concentration is markedly increased and the P_{CO_2} slightly decreased; the blood P_{O_2} is reduced. The results are compatible with a metabolic alkalosis due to vomiting caused by pyloric stenosis (see 14.1.12; p.**156**). The capacity for respiratory compensation, by decreasing the rate of ventilation in order to increase the P_{CO_2}, is limited because a reduction in the P_{O_2} or a rise in the P_{CO_2} stimulates the respiratory drive.

Diagnosis: a partially compensated metabolic alkalosis.

14.2.8 *Comment:* these findings are compatible with a partially compensated respiratory alkalosis; the slightly low blood actual bicarbonate concentration suggests that there is some metabolic compensation present. The low P_{O_2} may be caused by a ventilation/perfusion mismatch, possibly due to pulmonary collapse. Ventilator settings must be adjusted to ensure adequate oxygenation; carbon dioxide diffuses much more rapidly across alveolar cell membranes than oxygen, even if there is pulmonary oedema. The fall in P_{CO_2} slows the carbonate dehydratase mechanism in the renal tubular cells and erythrocytes and consequently the blood actual bicarbonate concentration falls; however, the ability to compensate for a respiratory alkalosis is limited.

Diagnosis: a partially compensated respiratory alkalosis.

14.2.9 *Comment:* the very low blood pH and actual bicarbonate concentration is compatible with a marked metabolic acidosis, caused by diabetic ketoacidosis in this patient. The P_{CO_2} is reduced indicating partial respiratory compensation due the hyperventilation with a fall in the blood P_{CO_2} as carbon dioxide is 'blown off'. This is characterized clinically by deep sighing respiration ('Kussmaul' respiration).

Diagnosis: partially compensated metabolic acidosis secondary to diabetic ketoacidosis.

14.2.10 *Comment:* these results are compatible with a partially compensated metabolic acidosis. A urinary pH of more than 5.5 is inappropriately high in the presence of a low blood pH; this finding suggests that there is impaired renal tubular hydrogen ion excretion, possibly due to renal tubular acidosis. However, an alternative and more usual explanation could be a urinary tract infection with an organism such as *Proteus vulgaris* which, by converting urea to ammonia, makes the urine alkaline *in vitro*. However, if this is the cause the urinary pH is usually much higher. The urinary pH should be measured on a fresh, and if possible, sterile urinary specimen. The measurement of blood gases is not necessary to make this diagnosis, but the results have been included here as an illustration.

Diagnosis: renal tubular acidosis.

14.3 DISORDERS OF CALCIUM AND PHOSPHATE METABOLISM

14.3.1 A 76-year-old woman presented with inoperable breast carcinoma.

Plasma		Reference range
Calcium	3.94 mmol/L	2.15–2.55
Albumin	36 g/L	27–42
Phosphate	1.01 mmol/L	0.60–1.40
Urea	10.1 mmol/L	2.5–7.0

14.3.2 A 66-year-old woman, on long-term vitamin D treatment following a parathyroidectomy, attended for a 'routine' medical examination.

Plasma		Reference range
Calcium	2.95 mmol/L	2.15–2.60
Albumin	31 g/L	30–45
Phosphate	1.91 mmol/L	0.60–1.40
Urea	8.1 mmol/L	2.5–7.0

14.3.3 A 30-year-old man was referred for investigation of renal calculi.

Plasma		Reference range
Calcium	2.80 mmol/L	2.15–2.55
Albumin	36 g/L	27–42
Phosphate	0.55 mmol/L	0.60–1.40
Urea	3.3 mmol/L	2.5–7.0

14.3.4 A 64-year-old man had the following investigations performed immediately after a coronary artery bypass graft.

Plasma		Reference range
Calcium	2.76 mmol/L	2.15–2.60
Albumin	31 g/L	30–45

14.3.1 *Interpretation:* severe hypercalcaemia with a relatively low plasma phosphate concentration for the reduced GFR indicated by the increased plasma urea concentration.

Comment: the combination of hypercalcaemia and a plasma phosphate concentration which, although within the reference range, is inappropriately low for the reduced GFR, is suggestive of either primary or tertiary hyperparathyroidism, or humoral hypercalcaemia of malignancy (p.**82**). The severity of the hypercalcaemia and the clinical findings suggest that humoral hypercalcaemia of malignancy is the most likely diagnosis.

Diagnosis: humoral hypercalcaemia of malignancy.

14.3.2 *Interpretation:* hypercalcaemia with hyperphosphataemia.

Comment: hypercalcaemia with hyperphosphataemia usually occurs when there is increased calcium and phosphate absorption from the gastrointestinal tract or reabsorption from bone. This is most commonly caused by increased vitamin D activity, either due to iatrogenic vitamin D overdosage, or to sarcoid or other granulomatous disorders (p.**82**). Less common causes include extensive bony metastases or thyrotoxicosis (both very rare).

Diagnosis: hypervitaminosis D causing hypercalcaemia in a patient receiving treatment with calciferol following a parathyroidectomy.

14.3.3 *Interpretation:* mild hypercalcaemia and hypophosphataemia.

Comment: the combination of hypercalcaemia and hypophosphataemia in the absence of any significant medical history is suggestive of either primary hyperparathyroidism or humoral hypercalcaemia of malignancy. Because the plasma calcium concentration is only marginally raised and because the presence of renal calculi implies that the disorder has been present for some time, primary hyperparathyroidism is the more likely diagnosis.

Diagnosis: primary hyperparathyroidism due to a parathyroid adenoma.

14.3.4 *Interpretation:* mild hypercalcaemia.

Comment: mild hypercalcaemia may be documented during or immediately after surgery when large volumes of blood products containing calcium and citrate have been transfused. The concentration of the physiologically active free ionized fraction of calcium may be often low due to the chelation of calcium ions by citrate. As citrate is metabolized in the liver the concentration of free ionized calcium in the blood rises.

Diagnosis: post-transfusion hypercalcaemia.

14.4 DISORDERS OF CARBOHYDRATE METABOLISM – INTERPRETATION OF GLUCOSE LOAD (TOLERANCE) TEST

14.4.1 A 35-year-old man presented with recurrent boils; a random plasma glucose concentration was 6.8 mmol/L. A glucose load (tolerance) test was performed.

Time (hours)	Plasma glucose (mmol/L)	Urine glucose	Ketones
0	4.5	negative	+
1	8.4	negative	negative
2	5.5	negative	negative

14.4.2 A 73-year-old man was found to have glycosuria on routine investigation at the time of admission to hospital for an elective operation. A glucose load (tolerance) test was performed.

Time (hours)	Plasma glucose (mmol/L)	Urine glucose	Ketones
0	6.3	negative	negative
1	10.7	++	negative
2	14.0	+++	negative

14.4.3 A 24-year-old woman had a glucose load (tolerance) test performed at her first antenatal clinic visit as she had previously given birth to a large infant weighing 4.2 kg.

Time (hours)	Plasma glucose (mmol/L)	Urine glucose	Ketones
0	4.2	negative	negative
1	9.4	++	negative
2	7.5	++	negative

14.4.1 *Interpretation:* both the fasting and the two-hour plasma glucose concentrations were normal. There was starvation-induced ketosis present in the first urine sample; this is a relatively common finding.

Comment: recurrent episodes of skin sepsis with boils is one of the presenting features of diabetes mellitus. A normal random plasma glucose concentration of 6.8 mmol/L does not exclude the diagnosis of diabetes mellitus or impaired glucose tolerance and, therefore, a glucose load (tolerance) test was performed.

Diagnosis: normal glucose load (tolerance) test.

14.4.2 *Interpretation:* the initial fasting plasma glucose concentration was marginally increased but the two-hour result was within the range diagnostic of diabetes mellitus. Glycosuria was present when the plasma glucose concentration was 10.7 mmol/L.

Comment: the diagnosis of diabetes mellitus should only be made if the raised plasma glucose concentration at two hours is confirmed by an additional abnormal plasma glucose concentration, on either a random or two-hour postprandial sample.

Diagnosis: compatible with diabetes mellitus but the diagnosis needs to be confirmed by repeating the test on a different occasion.

14.4.3 *Interpretation:* a normal glucose load (tolerance) test.

Comment: although the glucose load (tolerance) test was normal glycosuria was present when the recorded plasma glucose concentration was only 9.4 mmol/L indicating a low renal threshold for glucose excretion (glycosuria). This finding may occur when the GFR increases, as during pregnancy.

Diagnosis: renal glycosuria.

14.5   DISORDERS OF THYROID FUNCTION

14.5.1 A patient who had previously had a thyroidectomy attended her general practitioner for a routine 'check up'.

Plasma			Reference range
Total T_4	<25	nmol/L	65–145
Free T_3	10.1	pmol/L	4.6–9.2
TSH	<0.2	mU/L	0.2–4.0

14.5.2 A 31-year-old patient presented with tiredness; he has epilepsy.

Plasma			Reference range
Total T_4	49	nmol/L	65–145
Free T_3	6.1	pmol/L	4.6–9.2
TSH	1.2	mU/L	0.2–4.0

14.5.3 A two-week-old child presented with prolonged unconjugated jaundice.

Plasma			Reference range
Total T_4	198	nmol/L	90–180
TSH	1.9	mU/L	0.2–4.0

14.5.4 A 27-year-old woman complained of weight gain of recent onset.

Plasma			Reference range
Total T_4	230	nmol/L	65–145
TSH	1.5	mU/L	0.2–4.0
TBG	26.6	mg/L	6.0–16.0

14.5.1 *Interpretation:* the undetectable plasma TSH concentration suggests increased negative feedback suppression of pituitary TSH release, consistent with the slightly high plasma free T_3 concentration. The plasma total T_4 concentration is very low.

Comment: the very low plasma total T_4 concentration was too low to be caused by low binding-protein concentrations; this is an uncommon finding in the presence of a slightly raised plasma free T_3 concentration. The most likely explanation was that the patient was on thyroid hormone replacement with liothyronine (T_3) rather than L-thyroxine (T_4).

Diagnosis: the patient was taking liothyronine.

14.5.2 *Interpretation:* normal plasma TSH and free T_3 concentrations suggest that the patient was biochemically euthyroid.

Comment: the combination of a low plasma total T_4 and a normal TSH concentration is most commonly due to a reduced thyroxine-binding globulin (TBG) concentration. Other possible explanations include the effects of anticonvulsant drugs, such as phenytoin, which stimulate 5'-deiodinase activity and increase the rate of conversion of T_4 to T_3. Other drugs, such as salicylates, bind to TBG and displace T_4, thus reducing the plasma total T_4 concentration. The plasma free T_4 and TSH concentrations remain normal.

Diagnosis: euthyroid patient on treatment with phenytoin.

14.5.3 *Interpretation:* the plasma total T_4 concentration is considerably increased but the plasma TSH concentration is normal.

Comment: hypothyroidism should be excluded in all infants who present with prolonged 'physiological' jaundice. Plasma T_4 concentrations are higher in newborns than in adults and then gradually fall. The plasma TSH concentration peaks within 24 hours of birth and falls to within the adult range within three to four days.

Diagnosis: normal thyroid function tests in a euthyroid infant.

14.5.4 *Interpretation:* the plasma total T_4 and TBG concentrations are increased; that of TSH is normal.

Comment: a raised plasma thyroxine-binding globulin (TBG) concentration causes an increase in the bound, physiologically inactive T_4 and T_3 concentrations, the commonest cause of which is due to the effect of the oestrogen containing contraceptive pill. The normal plasma TSH concentration confirms that this patient is euthyroid.

Diagnosis: oestrogen-induced increase in plasma TBG concentration.

14.5.5 A 39-year-old woman was on treatment with lithium carbonate for depression.

Plasma			Reference range
Total T$_4$	72	nmol/L	65–145
TSH	25.8	mU/L	0.2–4.0

14.5.6 A 25-year-old woman presented with palpitations; she had proptosis.

Plasma			Reference range
Free T$_4$	51.4	pmol/L	11.0–25.0
Free T$_3$	35.6	pmol/L	4.6–9.2
TSH	<0.2	mU/L	0.2–4.0

14.5.5 *Interpretation:* the plasma total T_4 concentration is low-normal and the plasma TSH concentration increased.

Comment: these results indicate borderline primary hypothyroidism, a slight fall in T_4 reducing the negative feedback inhibition on TSH release from the anterior pituitary gland, thus resulting in an increase in TSH release and in the plasma concentration. The increased stimulation of the thyroid gland partially or completely restores the thyroxine production rate, and therefore plasma T_4 concentration, towards normal. Lithium carbonate treatment contributes to the development of hypothyroidism as it inhibits the release of thyroxine from the thyroid gland.

Diagnosis: lithium-induced early primary hypothyroidism.

14.5.6 *Interpretation:* the plasma free T_4 and T_3 concentrations are very high and are associated with an undetectable plasma TSH concentration diagnostic of hyper-thyroidism.

Comment: these findings are typical for a patient presenting with thyrotoxicosis. The measurement of the plasma concentrations of both free T_4 and T_3 was unnecessary; in most cases the diagnosis can be confirmed by measuring the plasma total or free T_4 and TSH concentrations only.

Diagnosis: Graves' disease (autoimmune thyrotoxicosis).

Investigations

15 Investigations

15.1 INVESTIGATION OF RENAL GLOMERULAR AND TUBULAR FUNCTIONS

Creatinine clearance

Indication: This investigation is rarely useful. If the plasma creatinine and urea concentrations are unequivocally high these simple investigations can be used to monitor progress. Renal function is marginally impaired in many, especially elderly, people and, unless progressive, is compatible with normal life. Such mild impairment cannot and need not be treated. Very occasionally estimation of the glomerular filtration rate may be indicated if nephrotoxic drugs are to be given, particularly if minimal or mild glomerular dysfunction is suspected.

Procedure:

- *Urine:* collect a sample of urine over a known period of time, usually 24 hours. Ask the patient to empty his bladder and discard the urine; record the time, for example 07.00 hours on day 1. Collect all the urine passed, up to and including a sample at 07.00 hours on day 2, into the container provided by the laboratory. This container must be brought to the laboratory as soon as possible after the collection has been completed. The patient *must* be informed if a preservative has been added to the container before the collection has started.
- *Blood:* collect a sample of blood for the determination of plasma creatinine concentration. Ideally this sample should be taken in the middle of the urine collection but for practical purposes it is normally taken just before or just after the urine collection.

Calculation:

$$\text{Creatinine clearance (ml/min)} = \frac{\text{urinary [creatinine]} \times \text{urine volume (ml)}}{\text{plasma [creatinine]} \times \text{collection time (min)}}$$

Ammonium chloride loading test

Indication: the diagnosis of distal renal tubular disorders. This test is not necessary if the pH of a urine specimen, collected overnight, is less than 5.5 or above 5.5 when the patient is frankly acidotic. Ammonium chloride is potentially acid because it dissociates to ammonia and H^+ ions. After ingestion the kidneys usually secrete the H^+ and the urinary pH falls.

Procedure: no food and fluid is taken after midnight.

- 08.00 hours ammonium chloride in gelatine capsules is given orally at a dose of 0.1 g/kg body weight;
- urine specimens are collected hourly for up to eight hours and the pH of each is measured immediately *by the laboratory.*

Interpretation: in normal subjects the urine pH falls to 5.5 or below at between two and eight hours after the dose. In generalized tubular disorders the response of the functioning nephrons may give a normal result. In distal renal tubular acidosis, this degree of acidification does not occur. Urinary acidification is normal in proximal renal tubular acidosis.

Water deprivation test

Indication: the diagnosis of diabetes insipidus (DI). The restriction of oral water intake for some hours should stimulate ADH secretion. Solute-free water is reabsorbed from the renal collecting ducts and a concentrated urine is passed. Maximal water reabsorption is impaired if either:

- ADH release from the posterior pituitary gland is impaired;
- the renal collecting tubules are unable to respond normally with increased permeability to water in the presence of ADH;
- the countercurrent multiplication mechanism in the loops of Henle is impaired.

If the control of ADH release is impaired (cranial DI) and renal tubular function is normal, the administration of ADH will increase the renal concentrating ability and increase the urinary osmolality. If renal tubular responsiveness to ADH is impaired (nephrogenic DI) plasma ADH concentrations will be high and administration of additional hormone will not improve the renal concentrating power. The test must be stopped if:

- the patient becomes clinically distressed;
- the plasma osmolality (or sodium concentration) rises to high or high-normal levels. Urine and blood specimens should be collected at once;
- the patient loses more than three per cent of his body weight.

Procedure: always contact the laboratory *before* starting the test, both to ensure efficient and speedy analysis and to check local variations in the protocol.

The patient is allowed no food or water after about 20.00 hours on the night before the test. *He must be in hospital and kept under observation* during the period of fluid restriction. The patient should be weighed before the test is started. *The duration of water deprivation depends on the clinical presentation and the degree of polyuria.*

On the day of the test at:

- 08.00 hours the bladder is emptied. Blood and urine are collected and plasma and urinary osmolalities measured. If the plasma osmolality is:
    low-normal or low, water depletion is unlikely and polyuria is probably due to an appropriate response to a high intake;
    high-normal or high, and the urinary osmolality is greater than 850 mmol/kg, the test can be stopped as this excludes the diagnosis of DI;
    high and the urinary osmolality is low this is diagnostic of DI and further fluid restriction is not necessary and may be dangerous.
- hourly intervals blood and urine are again collected and the plasma and urinary osmolality measured. Continue until at least three consecutive samples show no increase in urinary osmolality (proceed to a DDAVP test) or it exceeds 850 mmol/kg.

Interpretation:

- If the urinary osmolality exceeds 850 mmol/kg neither significant tubular disease nor DI are present and the test may be stopped.
- If the urinary osmolality is below 850 mmol/kg, and that of plasma is normal, fluid restriction should be continued and the estimations repeated at hourly intervals. The test should be stopped as soon as the urinary osmolality exceeds 850 mmol/kg.

- Failure to concentrate three consecutive urine specimens indicates either tubular disease or DI; the differential diagnosis is usually clear on clinical grounds and the test may then be stopped. If the diagnosis of DI is considered, continue with the DDAVP test.

<u>Warning</u>. *This test should not be performed if the patient is volume depleted or has even mild hypernatraemia.* In such cases the finding of a low urinary to plasma osmolality ratio is diagnostic and further fluid restriction may be dangerous.

DDAVP test

Indication: to distinguish between cranial and nephrogenic DI. DDAVP (1-**D**eamino 8-**D**-**a**rginine **v**aso**p**ressin; desmopressin acetate) is a potent synthetic analogue of vasopressin.

Procedure: 4 µg DDAVP is injected intramuscularly; blood and urine samples are collected hourly and the plasma and urinary osmolalities measured.

Interpretation: a diagnosis of DI must have been made following a water deprivation test.

- If there is tubular dysfunction there will be no change in the urine osmolality but the plasma osmolality may increase.
- In cranial diabetes insipidus the urine osmolality increases usually to a value greater than 850 mmol/kg.

<u>Note</u>. After prolonged overhydration, usually due to hysterical polydipsia, urinary concentrating ability, even after the administration of exogenous ADH, may be impaired. This is due to the 'washing out' of medullary hyperosmolality. Unless the patient is volume depleted or there is plasma hyperosmolality, the test should be repeated after several days of relative water restriction. The patient should be kept under careful observation, on the one hand for signs of genuine distress associated with a rise in plasma osmolality, and on the other for surreptitious drinking.

15.2 INVESTIGATION OF SUSPECTED DIABETES MELLITUS

Diabetes mellitus (DM) should not be diagnosed unless unequivocally high plasma glucose concentrations have been found in specimens taken on at least *two different occasions*. If the plasma glucose concentrations are above the reference limits but are below those defined for DM, the patient is said to have impaired glucose tolerance.

Blood for glucose estimation should be taken if a patient presents with symptoms that can be attributed to DM, has glycosuria, or if it is desirable to exclude the diagnosis because, for example, of a strong family history. Blood samples may be taken:

- after at least 10 hours of fasting;
- two hours after a meal;

Table 15.1 Interpretation of fasting and random plasma glucose concentrations

| | Venous plasma glucose (mmol/L) | |
	Fasting	Random
Diabetes unlikely	5.5 or less	
Diabetic	7.8 or more	11.1 or more

- at random;
- as part of a glucose tolerance test.

Diabetes mellitus is confirmed if *one* of the following is present:

- a fasting plasma glucose concentration of 7.8 mmol/L or more *on two occasions*;
- a random plasma glucose concentration of 11.1 mmol/L or more *on two occasions*;
- both a fasting plasma glucose concentration of more than 7.8 mmol/L and a random concentration of more than 11.1 mmol/L.

Diabetes mellitus is usually excluded if:

- a fasting plasma glucose concentration is less than 5.5 mmol/L on two occasions. Samples taken at random times after meals are less reliable for excluding than for confirming the diagnosis.

These diagnostic limits are summarized in Table 15.1. Indications for performing an oral glucose tolerance test are rare. They include:

- a fasting plasma glucose concentrations between 5.5 and 7.8 mmol/L;
- a random plasma glucose concentration between 7.8 and 11.1 mmol/L;
- a high index of clinical suspicion.

A patient without symptoms may be suspected of having diabetes mellitus because of the chance finding of glycosuria or hyperglycaemia. If the plasma glucose concentration is more than 11.1 mmol/L two hours after a standard glucose load the finding should be confirmed before a definitive diagnosis is made. Transient hyperglycaemia may occur as a response to acute metabolic stress such as major surgery or after a myocardial infarct; the investigations for DM should be delayed until recovery.

If the glucose concentration is measured in whole blood the concentrations will be approximately 1.0 mmol/L lower.

Oral glucose load (tolerance) test (GTT)

Procedure: the patient should be resting and should not smoke during the test.

- The patient fasts overnight (for at least 10, but not more than 16, hours). Water, but no other beverage, is allowed.
- A venous sample is withdrawn for plasma glucose estimation. A urine specimen is collected using the double voided technique.
- A solution containing 75 g of glucose and made up to approximately 300 ml with water is drunk slowly over about five minutes. This hyperosmolar solution may cause nausea and occasionally vomiting and diarrhoea, with delayed glucose absorption. It is, therefore, more usual to give a solution of a mixture

Table 15.2 Frequency of sampling and duration of test, depending on the indication for performing a glucose load (tolerance) test

Diagnosis	Frequency of sampling (minutes)	Duration of test (hours)
Diabetes mellitus	120	2
Renal threshold for glucose	60	2
Reactive hypoglycaemia	30	5
Excess growth hormone	30	2

Table 15.3 Interpretation of a 75 g oral glucose load (tolerance) test

	Venous plasma glucose (mmol/L) Fasting	Two hours
Diabetes unlikely	5.5 or less	7.8 or less
Impaired glucose tolerance	5.5–7.8	7.8–11.1
Diabetic	7.8 or more	11.1 or more

of glucose and its oligosaccharides, as fewer molecules have less of an osmotic effect than the equivalent amount of monosaccharide; the oligosaccharides are all hydrolysed at the brush border, and the glucose immediately enters the cells. Solutions which contain the equivalent of 75 g of anhydrous glucose include:

    113 ml of 'Fortical' (Cow and Gate Ltd);

    353 ml of 'Lucozade' (SmithKline Beecham Ltd).

- Further blood and urine samples are taken at two hours after the ingestion of glucose. The frequency of blood sampling is different depending on the indications for performing the investigation. Examples are shown in Table 15.2.

The plasma glucose concentrations are measured and the urine samples tested for glucose and ketones.

Interpretation of the oral glucose load (tolerance) test is shown in Table 15.3. The following factors may affect the results:

- *Previous diet*. No special restrictions are necessary if the patient has been on a normal diet for three to four days. However, if the test is performed after a period of carbohydrate restriction, perhaps as part of a reducing diet, this may cause abnormal glucose tolerance, probably because metabolism is adjusted to the 'fasted' state and so favours gluconeogenesis.
- *Time of day*. Most glucose tolerance tests are performed in the morning and the reference values quoted are for this time of day. There is evidence that tests performed in the afternoon yield higher plasma glucose concentrations and that the accepted 'reference values' may not be applicable. This may be due to a circadian variation in islet cell responsiveness.
- *Drugs*. Drugs such as steroids, oral contraceptives and thiazide and loop diuretics may impair glucose tolerance.
- *Pregnancy*. The interpretation is the same for pregnant as for nonpregnant women.

15.3 PROCEDURE FOR REPEATED BLOOD SAMPLING

Many 'dynamic' tests of endocrine function require several blood samples to be taken over a short period of time. Repeated venepuncture is unpleasant for the patient and is also undesirable because it may cause stress, and so interfere with the results. Insertion of an indwelling needle or cannula helps to minimize these problems and enables intravenous therapy to be given without delay if there should be an untoward reaction such as hypoglycaemia. The following procedure is recommended:

- A needle or cannula, at least as large as a 19G, is inserted into a vein at a site that preferably will not be subjected to movement or flexion and secured in position with adhesive strapping.
- Isotonic saline is infused slowly or the cannula flushed with sodium citrate, to keep the needle open. Heparin should not be used because it interferes with some assays.
- The first specimen should not be taken for at least 30 minutes, so that any stress-induced elevation of hormones may decrease to basal concentrations. All samples may be taken through this needle as follows:
    disconnect the saline infusion, or use a three-way tap between the infusion set and the needle;
    aspirate and discard about 2 ml of the saline or citrate blood mixture;
    using a different syringe aspirate the specimen for the hormone assay;
    reconnect and restart the saline flow.

15.4 INVESTIGATION OF ENDOCRINE DISORDERS

Combined pituitary stimulation test

This test is potentially dangerous and must be done under direct medical supervision. It is contraindicated in patients with:

- ischaemic heart disease;
- epilepsy.

Glucose for intravenous administration must be immediately available in case severe hypoglycaemia develops. If infused, care must be taken that patients, particularly children, do not develop hyperglycaemia. *After the test the patient must be given something to eat.*

The plasma concentrations of anterior pituitary hormones are measured after stimulation by stress, TRH and GnRH. Plasma cortisol concentrations are usually measured as an index of ACTH secretion; the entire hypothalamic–pituitary target gland axis is therefore tested. If glucose needs to be given, continue with the sampling. The stress has certainly been adequate to stimulate hormone secretion.

Procedure: after an overnight fast:

- insert an indwelling intravenous cannula;
- after at least 30 minutes, take basal samples;
- inject soluble insulin in a high enough dose to lower the plasma glucose concentration to less than 2.5 mmol/L and to cause symptomatic hypoglycaemia. The recommended dose of insulin must be adjusted for the patient's body weight (BW) and for the suspected clinical condition under investigation:

    The usual dose is 0.15 U/kg BW.

    If pituitary or adrenocortical hypofunction are suspected, or if a low fasting plasma glucose concentration has been found, reduce the dose to 0.1 or 0.05 U/kg BW.

    If there is resistance to the action of insulin because of Cushing's syndrome, acromegaly or obesity, 0.2 or 0.4 U/kg BW may be needed.

- immediately inject 200 µg of TRH and 100 µg of GnRH;
- take blood samples at least 30, 45, 60, 90 and 120 minutes after the injections and request hormone assays as indicated in Table 15.4.

Interpretation: methods of hormone assay vary and results must be compared with reference values issued by the same laboratory. The following is intended as a guide only.

If hypoglycaemia has been adequate:

- plasma cortisol concentrations should rise by more than 200 nmol/L and exceed 550 nmol/L;
- plasma GH concentrations should exceed 20 mU/L;
- plasma TSH concentrations increase by at least 2 mU/L and exceed the upper limit of the reference range;
- plasma prolactin and gonadotrophin concentrations should rise significantly for the reference range for the method.

If the plasma cortisol concentration does not rise a tetracosactrin ('Synacthen') test should be performed to exclude primary adrenocortical hypofunction.

Table 15.4 Protocol for a combined pituitary stimulation test

	Time (minutes) following injection					
	-10	0	30	60	90	120
Insulin						
glucose	√		√	√	√	√
cortisol	√		√	√	√	√
growth hormone	√		√	√	√	√
TRH						
T₄		√				
TSH		√	√	√		
prolactin		√	√	√		
GnRH						
LH		√	√	√		
FSH		√	√	√		

√ indicates when blood should be analysed for specific hormones

Gonadotrophin-releasing hormone (GnRH) test

Synthetic GnRH stimulates the release of gonadotrophins (LH and FSH) from the normal anterior pituitary gland.

Procedure:

- 100 μg of GnRH is given by rapid intravenous injection.
- Plasma LH and FSH concentrations are measured in blood samples drawn before and at 30 and 60 minutes after the injection.

Interpretation: in normal subjects the plasma LH concentration rises by at least 5 U/L; this rise fails to occur in patients with pituitary hypofunction.

Thyrotrophin-releasing hormone (TRH) test

The TRH test may be used to confirm the diagnosis of secondary hypothyroidism, or occasionally to diagnose early primary hypothyroidism.

Procedure:

- 200 μg of TRH in 2 ml saline is injected intravenously over about a minute.
- Plasma TSH concentration is measured on blood samples drawn before and at 20 and 60 minutes after the injection.

Interpretation: in normal subjects the plasma TSH concentration increases by at least 2 mU/L and exceeds the upper limit of the reference range. The maximum response occurs at 20 minutes.

- A high-normal basal plasma TSH concentration, with an exaggerated response at 20 minutes and a slight fall at 60 minutes is suggestive of early primary hypothyroidism.
- A subnormal rise of TSH confirms the diagnosis of secondary hypothyroidism of pituitary origin.
- A normal or an exaggerated but delayed rise, with plasma TSH concentrations higher at 60 minutes than at 20 minutes, suggests secondary hypothyroidism due to hypothalamic dysfunction. If clinically indicated, pituitary and hypothalamic function should be investigated.
- A flat response is compatible with hyperthyroidism.

Investigation of suspected Cushing's syndrome

Because of the serious nature of Cushing's syndrome, and because it may be treatable, the diagnosis must be excluded even when clinical features are only suggestive. Initial tests may exclude the diagnosis, but may yield some 'false positive' results.

HAS THE PATIENT GOT CUSHING'S SYNDROME?

The following tests can be carried out initially without admitting the patient into hospital:

- *Overnight dexamethasone suppression test.* Dexamethasone (2 mg) is given as a single oral dose at 23.00 hours. The plasma cortisol concentration is measured on a specimen taken at 09.00 hours the next morning. Suppression is defined as a plasma cortisol concentration of less than 190 nmol/L.
- *Twenty-four-hour urinary free cortisol estimation.* Cortisol is assayed on a 24-hour urine collection. An increased excretion is suggestive of Cushing's syndrome.

If these tests are normal, it is unlikely that the patient has Cushing's syndrome. If either is abnormal, or if there is a strong clinical suspicion, further tests should be carried out in hospital.

- *Loss of the circadian rhythm of plasma cortisol* may be demonstrated by measuring the plasma cortisol concentration on blood samples collected at 09.00 hours and at 23.00 hours. This test can only really be performed in hospital and may be followed by repeating the overnight dexamethasone test.

If all these tests are normal, follow-up the patient in the out-patient clinic. The manifestations of Cushing's syndrome may be intermittent and tests may have to be repeated later.

If the results show abnormal cortisol secretion consider alternative causes such as:

- stress, even that associated with admission to hospital;
- chronic excess alcohol ingestion;
- endogenous depression.

If the latter is likely, perform an insulin stimulation test. A normal cortisol response to this suggests endogenous depression rather than Cushing's syndrome.

Once the diagnosis of Cushing's syndrome has been made, proceed to the next step.

INVESTIGATE THE CAUSE OF CUSHING'S SYNDROME

Extremely high plasma cortisol or urinary free cortisol concentrations are suggestive of either adrenocortical carcinoma, especially if the patient is virilized, or of ectopic ACTH production. Investigate clinically and using other investigations, bearing in mind the possibility of carcinoma of the adrenal gland or bronchus. Severe hypokalaemic alkalosis suggests ectopic ACTH secretion.

Estimate plasma ACTH concentrations

Moderately raised plasma ACTH concentrations, even when within the upper limit of the reference range, are suggestive of either Cushing's disease or occult ectopic ACTH secretion. Very high plasma concentrations are found with overt ectopic ACTH secreting tumours, whereas in patients with adrenocortical tumours the plasma ACTH concentration is low. Remember that stress increases the plasma ACTH and cortisol concentrations.

High-dose dexamethasone suppression test

This test may distinguish between Cushing's disease and adrenocortical tumour or overt ectopic ACTH production.

Procedure: dexamethasone (2 mg) is given orally *every six hours* for two days, starting at 09.00 hours. Plasma cortisol is measured in a specimen taken at 09.00 hours on the first and third days.

Interpretation: suppression is defined as plasma cortisol concentrations less than 50 per cent of previously measured concentrations.

- Suppression suggests either pituitary-dependent Cushing's disease or an occult ACTH secreting tumour.
- Failure to suppress suggests either an adrenocortical tumour (low plasma ACTH concentrations) or overt ectopic ACTH production (very high plasma ACTH concentrations).

Anticonvulsant drugs, particularly phenytoin, may interfere with the dexamethasone suppression tests. They induce liver enzymes that increase the rate of metabolism of dexamethasone. Plasma concentrations may, therefore, be too low to cause negative feedback suppression.

Investigation of suspected adrenal hypofunction

SUSPECTED ADDISONIAN CRISIS

Procedure: take blood for immediate estimation of plasma urea and electrolyte concentrations, and for plasma cortisol, which can be stored and analysed later. Consider performing a short tetracosactrin ('Synacthen') stimulation test and start steroid treatment. Do not wait for the results of the laboratory tests.

Interpretation: hyponatraemia, hyperkalaemia and uraemia, although compatible with an Addisonian crisis, are common in many clinically similar acute conditions. Treat appropriately. The plasma cortisol concentration may be estimated later:

- if it is very high an Addisonian crisis is excluded;
- if is very low or undetectable, and if there is no reason to suspect CBG deficiency, an Addisonian crisis is confirmed.

Plasma cortisol concentrations, which would be 'normal' under basal conditions, may be inappropriately low for the degree of stress. If the result is equivocal perform a short tetracosactrin ('Synacthen') test.

Tetracosactrin ('Synacthen') test for investigation of adrenal hypofunction

Tetracosactrin is marketed as 'Synacthen' (Ciba).

Short tetracosactrin stimulation test

Procedure: the patient should be resting quietly.

- 250 µg of 'Synacthen', dissolved in about 1 ml of sterile water or isotonic saline, is given by intramuscular injection.

- Plasma cortisol concentrations are measured in blood samples drawn before and at 30 and 45 minutes after the injection.

Interpretation: normally the plasma cortisol concentration increases by at least 200 nmol/L, to a concentration of at least 550 nmol/L.

Five-hour tetracosactrin stimulation test

Indication: investigation of suspected chronic adrenal insufficiency

Procedure:

- 1 mg of depot tetracosactrin is injected intramuscularly;
- Plasma cortisol concentrations are measured in blood samples drawn before and at one and five hours after the injection.

Interpretation: normally plasma cortisol concentration rises to between 600 and 1300 nmol/L at one hour and to between 1000 and 1800 nmol/L at five hours.

Three-day tetracosactrin stimulation test

Indication: investigation of suspected chronic adrenal insufficiency if the results of a five-hour tetracosactrin stimulation test are equivocal.

Procedure:

- 1 mg of depot tetracosactrin is injected intramuscularly each day for three days
- Plasma cortisol concentrations are measured in blood samples drawn before and five hours after each injection.

Interpretation: normally plasma cortisol concentrations rise to between 1000 and 1800 nmol/L.

An increasing response to the short, five-hour and three-day tetracosactrin tests indicates gradual recovery of adrenal cortical function following prolonged lack of ACTH and suggests either hypothalamic or pituitary hypofunction. An impaired response to all these tests confirms primary adrenal hypofunction.

<u>Warning</u> Repeated injections of depot tetracosactrin may lead to sodium and water retention; this test is contraindicated in patients, such as those with congestive cardiac failure, in whom such retention may be dangerous.

Investigation of suspected growth hormone (GH) deficiency

A single high plasma GH concentration probably excludes deficiency. The initial sample should be taken when the highest physiological concentrations occur; immediately after exercise or during sleep. If plasma GH concentrations do not exclude deficiency, an additional independent test should be performed before GH deficiency is diagnosed.

Insulin stimulation test

This test is similar to the combined stimulation test but only insulin is given (see above).

Interpretation: plasma GH concentrations should rise and should exceed 20 mU/L

Investigation of suspected acromegaly or gigantism

A raised plasma GH concentration, which fails to suppress normally in response to a rising plasma glucose concentration, suggests autonomous hormone secretion. Basal plasma GH concentrations may be, but are not always, high enough to confirm the diagnosis. It saves time to start with a glucose suppression test.

Glucose suppression test

Procedure: after an overnight fast:

- insert an indwelling intravenous cannula.
- take basal samples for plasma glucose and GH estimation after at least 30 minutes.
- the patient drinks 75 g of glucose dissolved in 300 ml of water, or an equivalent glucose load.
- samples are taken for glucose and GH assay at 30, 60, 90 and 120 minutes after the glucose load has been taken.

Interpretation: in normal subjects, plasma GH concentrations fall to less than 2 mU/L. Although failure to suppress suggests acromegaly or gigantism, it may be found in some patients with severe liver or renal disease, in heroin addicts or in those taking levodopa.

The plasma glucose concentrations may demonstrate impaired glucose tolerance.

Investigation of suspected Conn's syndrome

Indication: investigation of a patient who presents with mild to moderate hypertension, a hypokalaemic alkalosis and a high-normal plasma sodium concentration and who is on no medication.

Procedure: the patient should be off all diuretics and antihypertensive drugs. He should be normovolaemic and have a daily sodium intake of about 100 mmol. After strict overnight (eight-hour) recumbency in hospital, blood is taken for the measurement of plasma electrolytes and for renin activity and aldosterone concentration. The patient must then walk around and further blood samples are taken 30 minutes and four hours later again for the measurement of plasma renin activity and aldosterone concentration.

Interpretation: the diagnosis of primary hyperaldosteronism is made if the plasma aldosterone concentration is high, and the plasma renin activity inappropriately low.

Plasma	Supine	Erect	
Aldosterone	100–450		pmol/L
Renin activity	1.1–2.7	2.8–4.5	pmol/ml/hour

Characteristically, in Conn's syndrome, the plasma renin activity remains suppressed even after assumption of the erect posture for 30 minutes. A fall in the plasma aldosterone concentration after four hours suggests that the cause is an adrenal adenoma rather than bilateral nodular hyperplasia.

Appendix

1

APPROXIMATE CONVERSION FACTORS FOR SI UNITS

	From SI units		To SI units	
Bilirubin	μmol/L $\times$ 0.058	= mg/dl	mg/dl $\div$ 0.058	= μmol/L
Calcium				
Plasma	mmol/L $\times$ 4	= mg/dl	mg/dl $\div$ 4	= mmol/L
Urine	mmol/24hours $\times$ 40	= mg/24 hours	mg/24hours $\div$ 40	= mmol/24hours
Cholesterol	mmol/L $\times$ 39	= mg/dl	mg/dl $\div$ 39	= mmol/L
Cortisol				
Plasma	nmol/L $\times$ 0.036	= μg/dl	μg/dl $\div$ 0.036	= nmol/L
Urine	nmol/24hours $\times$ 0.36	= μg/24hours	μg/24hours $\div$ 0.36	= nmol/24hours
Creatinine				
Plasma	μmol/L $\times$ 0.011	= mg/dl	mg/dl $\div$ 0.011	= μmol/L
Urine	μmol/24hours $\times$ 0.11	= mg/24hours	mg/24hours $\div$ 0.11	= μmol/24hours
PO_2	kPa $\times$ 7.5	= mmHg	mmHg $\div$ 7.5	= kPa
PCO_2	kPa $\times$ 7.5	= mmHg	mmHg $\div$ 7.5	= kPa
Glucose	mmol/L $\times$ 18	= mg/dl	mg/dl $\div$ 18	= mmol/L
Iron	μmol/L $\times$ 5.6	= μg/dl	μg/dl $\div$ 5.6	= μmol/L
TIBC	μmol/L $\times$ 5.6	= μg/dl	μg/dl $\div$ 5.6	= μmol/L
Phosphate	mmol/L $\times$ 3	= mg/dl	mg/dl $\div$ 3	= mmol/L
Proteins				
All serum	g/L $\div$ 10	= g/dl	g/dl $\times$ 10	= g/L
Urine	g/L $\times$ 100	= mg/dl	mg/dl $\div$ 100	= g/L
	g/24hours		No change	
Urate	mmol/L $\times$ 17	= mg/dl	mg/dl $\div$ 17	= mmol/L
Urea				
Plasma	mmol/L $\times$ 6	= mg/dl	mg/dl $\div$ 6	= mmol/L
Urine	mmol/24hours $\times$ 60	= mg/24hours	mg/24hours $\div$ 60	= mmol/24hours
5-HIAA	μmol/24hours $\times$ 0.2	= mg/24hours	mg/24hours $\div$ 0.2	= μmol/24hours
HMMA	μmol/24hours $\times$ 0.2	= mg/24hours	mg/24hours $\div$ 0.2	= mmol/24hours
Faecal 'fat'	mmol/24hours $\times$ 0.3	= g/24hours	g/24hours $\div$ 0.3	= mmol/24hours

A_ppendix_

REFERENCE RANGES

These reference ranges have been taken from one of the authors' laboratories. *They may differ quite considerably from those issued from the student's own laboratory*; those marked with an * indicate those which are most likely to have considerably different values in different laboratories (p.4)

Creatinine	μmol/L	60–120	male
		55–110	female
Urea	mmol/L	2.5–8.0	male
		2.5–7.0	female
Sodium	mmol/L	135–145	
Potassium	mmol/L	3.5–4.8	
Chloride	mmol/L	95–108	
$T\text{CO}_2$ (bicarbonate)	mmol/L	22–32	
Urate	mmol/L	0.17–0.44	
Calcium	mmol/L	2.15–2.55	
Phosphate	mmol/L	0.60–1.40	adult
		1.20–1.95	child 1–10 years
Magnesium	mmol/L	0.70–1.00	
Total protein	g/L	60–85	
Albumin	g/L	30–42	
Total bilirubin	μmol/L	<20	
Alanine transaminase*	U/L	5–40	
Aspartate transaminase*	U/L	5–40	
Alkaline phosphatase*	U/L	60–250	adult
		170–850	child 1–10 years
γ-glutamyltransferase*	U/L	5–55	male
		5–40	female
Lactate dehydrogenase* (hydroxybutyrate dehydrogenase*)	U/L	50–220	
Creatine kinase*	U/L	0–220	male
		0–150	female
Amylase*	U/L	90–300	total
		30–110	pancreatic
Glucose	mmol/L	2.2–5.5	fasting
HbA_{1c}	%	4.5–6.0	
Lipids			
total cholesterol	mmol/L	3.5–6.5	
triglyceride	mmol/L	0.5–2.2	
HDL-cholesterol	mmol/L	>0.9	
LDL-cholesterol	mmol/L	2.0–4.5	
Blood gases			
pH		7.35–7.45	
$P\text{CO}_2$	kPa	4.6–6.0	
$P\text{O}_2$	kPa	12–16	
Actual bicarbonate	mmol/L	22–30	

Iron	μmol/L	7–24	
Transferrin	g/L	2.0–3.2	
Total iron-binding capacity	μmol/L	49–78	
Ferritin	μg/L	10–400	male
		6–120	female

Endocrine tests

Thyroid function tests			
total T_4	nmol/L	65–145	
free T_4	pmol/L	11–25	
free T_3	pmol/L	4.6–9.2	
TSH	mU/L	0.2–4.0	
FSH	U/L	1.8–8.6	male
		4.1–9.5	female follicular phase
		10.6–45.0	menopause
LH	U/L	0.7–6.0	male
		1.5–11.5	female follicular phase
		8.5–53.0	menopause
Oestradiol	pmol/L	55–165	male
		110–183	female follicular phase
		<100	menopause
Progesterone	nmol/L	20–60	female luteal phase (day 20 to 22 of cycle)
Testosterone	nmol/L	13–30	male older than 18 years
		0–3	female older than 18 years
SHBG	nmol/L	20–45	male
		35–100	female
Cortisol	nmol/L	250–750	morning
		<250	midnight
Growth hormone	mU/L	<10	
		<2	GTT suppression test
Prolactin	mU/L	0–425	

Appendix

COMMON ABBREVIATIONS

ACE	Angiotensin-converting enzyme
ACP	Acid phosphatase
ACTH	Adrenocorticotrophic hormone (corticotrophin)
ADH	Antidiuretic hormone (arginine vasopressin; AVP)
AIDS	Acquired immune deficiency syndrome
ALP	Alkaline phosphatase
ALT	Alanine transaminase
Anti-HB$_c$	Antibody to hepatitis B viral core
Anti-HB$_s$	Antibody to hepatitis B surface antigen
AST	Aspartate transaminase
BJP	Bence Jones protein
CAH	Congenital adrenal hyperplasia
CK	Creatine kinase
CSF	Cerebrospinal fluid
DDAVP	1-Deamino-8-D-arginine vasopressin (desmopressin acetate)
ECF	Extracellular fluid
FSH	Follicle stimulating hormone
GFR	Glomerular filtration rate
GGT	γ-glutamyltransferase (γ-glutamyltranspeptidase; γ-GT)
GH	Growth hormone (somatotrophin)
GnRH	Gonadotrophin-releasing hormone
GTT	Glucose tolerance test
HAV	Hepatitis A virus
HBD	Hydroxybutyrate dehydrogenase
HB$_s$Ag	Hepatitis B surface antigen
HDL	High-density lipoprotein
hGH	Human growth hormone
HIV	Human immunodeficiency virus
HMMA	4-Hydroxy-3-methoxy-mandelate (VMA)
ICF	Intracellular fluid
IDDM	Insulin-dependent diabetes mellitus
LDL	Low-density lipoprotein
LH	Luteinizing hormone
MEN	Multiple endocrine neoplasia
MRI	Magnetic resonance imaging
NIDDM	Noninsulin-dependent diabetes mellitus
NSAID	Nonsteroidal anti-inflammatory drug
PBG	Porphobilinogen
PSA	Prostate specific antigen
PTH	Parathyroid hormone
PTHRP	Parathyroid hormone related protein
RTA	Renal tubular acidosis
SHBG	Sex-hormone-binding globulin
T$_3$	Tri-iodothyronine
T$_4$	Thyroxine
TBG	Thyroxine-binding globulin

TIBC	Total iron-binding capacity
TP	Total protein
TRH	Thyrotrophin-releasing hormone
TSH	Thyroid-stimulating hormone
VLDL	Very-low-density lipoprotein
VMA	Vanillyl mandelate (HMMA)
Z–E	
syndrome	Zollinger–Ellison syndrome

Index

Some of the page reference numbers are highlighted as follows:

- **bold** - main page reference with discussion;
- underlined - case presentation and data interpretation relating to subject;
- *italic* - subject mentioned in table only.